Peter Smith Michie

Elements of Analytical Mechanics

Fourth Edition

Peter Smith Michie

Elements of Analytical Mechanics
Fourth Edition

ISBN/EAN: 9783337277390

Printed in Europe, USA, Canada, Australia, Japan

Cover: Foto ©berggeist007 / pixelio.de

More available books at **www.hansebooks.com**

ELEMENTS

OF

ANALYTICAL MECHANICS.

BY

PETER S. MICHIE,

*Professor of Natural and Experimental Philosophy in the
U. S. Military Academy, West Point, and
Brevet Lieutenant-Colonel
U. S. Army.*

FOURTH EDITION.

SECOND THOUSAND.

NEW YORK:
JOHN WILEY & SONS.
LONDON: CHAPMAN & HALL, LIMITED.
1897.

ROBERT DRUMMOND, ELECTROTYPER AND PRINTER, NEW YORK.

PREFACE.

This volume is a revised edition of the text taught to the cadets of the U. S. Military Academy during the session of 1886-7. Together with a brief chapter on Hydromechanics, it is intended to comprise a four-months course of instruction for students well versed in elementary mathematics, including the Calculus.

The author has aimed to present a clear, concise, yet comprehensive course, covering all the important principles of Mechanics which form the basis of that scientific knowledge now required by the military profession. A thorough mastery of this volume will enable the student to comprehend, upon careful study, any of the more difficult works upon the same subject which his professional duties may require him to consult.

The table of contents gives in consecutive order a very full statement of the subjects discussed. The arrangement of the subject-matter and method of treatment adopted are in accord with the judgment of several able scientific officers who have been associated with the author in the instruction of cadets, and are the result of over sixteen years' experience in daily contact with bright students.

The subject-matter comprising the volume has been drawn from many sources, and modified to suit the requirements of the necessarily limited course. Those conversant with the subject will recognize that many of the articles and illustrative examples are taken from Price's Infinitesimal Calculus, Vols. III. and IV. The most prominent other sources are Poisson's Traité de Mécanique, Routh's Rigid Dynamics, and Levy's La Statique Graphique.

Lieut. William B. Gordon, Ordnance Department, U. S. Army, Assistant Professor of Philosophy U. S. Military Academy, is entitled to at least equal credit with the author for whatever may be found worthy of commendation in the book. Many of the demonstrations in the previous edition have been simplified by him, and in nearly every instance where a question has arisen the author has finally deferred to Lieut. Gordon's better judgment.

The author is also under great obligations to Lieut. Sidney E. Stuart, Ordnance Department, who has carefully gone over the work, and suggested important changes which, in most cases, have been adopted.

WEST POINT, N. Y., August, 1887.

CONTENTS.

MECHANICS OF SOLIDS.

Introductory.

ART.		PAGE
1.	Mechanics defined...	1
2.	Observation and experiment..	1

Matter.

3.	Elements, atom, molecule...	2
4.	Illustration of the distinction between elements and molecules........	2
5.	Mass..	2
6.	Density...	2
7.	Mass in terms of volume and density...............................	2
8.	Inertia..	3

Force.

9.	Force defined..	3
10.	Molecular forces...	3
11.	Elasticity...	4
12.	Gravitation...	4

Motion.

13.	Kinematics...	5
14.	Motion defined..	5
15.	Velocity; its measure; parallelopipedon of velocity................	6
16.	Acceleration; resultant acceleration................................	7
17.	Angular velocity and angular acceleration.........................	8

Physical Units.

18.	The British system; units of length, time, mass and force........	8
19.	Other units..	9
20.	The centimeter, gram, second system	9

Stresses and Motive Forces.

ART.		PAGE
21.	Statics and kinetics	10
22.	Measure of the intensity of a stress	10
23.	Measure of the intensity of a motive force	11
24.	Impulsive forces	12
25.	Action of forces on free bodies	13
26.	Representation of a force	14
27.	Rectangular components of a force	15
28.	Resultant of a system of forces	15
29.	The parallelogram of forces	17
30.	Application of the trigonometric functions to forces	17
31.	The resultant of forces with common point of application; coplanar forces	18
32.	The parallelopipedon of forces	19
33.	The moment of a force	20
34.	Representation of moments	21
35.	Component moments of a force	21
36.	Composition and resolution of moments	22
37.	The parallelopipedon of moments	23
38.	Parallel forces	24
39.	Two parallel forces	25
40.	Two forces whose action-lines are opposite; general conclusions for parallel forces	26
41.	A couple	26
42.	The moment of a couple	27

Gravity.

43.	Gravity; weight of a body	27
44.	Apparent weight of a body	28
45.	Molecular weights considered a system of parallel forces; weights proportional to masses	28
46.	Acceleration due to gravity	29
47.	Weight of the unit of mass	30
48.	The centre of gravity, the principle of the centre of mass; general formulas for the centre of gravity	31
49.	Determination of the centre of gravity; by symmetry	33
50.	Centre of gravity of lines; formulas for any line and any plane curve; examples, arcs of circle and cycloid	33
51.	Centre of gravity of surfaces; general formulas	35
52.	For plane surfaces; triangle, polygon, circular sector and segment, parabola and elliptic quadrant	36

ART.		PAGE
53.	Centre of gravity of surfaces of revolution; spherical zone and conical surface..	41
54.	Centre of gravity of volumes; general formulas	42
55.	Volumes of revolution; paraboloid and spheroid....	43
56.	Centre of gravity of volumes by single integration; a pyramid and cone	44
57.	Theorems of Pappus..	45

Graphical Statics.

58.	Introductory..	46
59.	Reciprocal figures and their conditions.............	46
60.	The force polygon...	47
61.	The polar polygon...	47
62.	The properties of the polar polygon.................................	49
63.	Problems in graphical statics: (1) general case; (2) simple Warren truss with equal loads; (3) the same with unequal loads; (4) other cases....	52

Work and Energy.

64.	Work done by a force..	60
65.	Graphical representation of work....................................	61
66.	Energy defined; kinetic and potential energy....	62
67.	The law of the conservation of energy...............................	64
68.	The principle of virtual velocities....................................	66
69, 70	The fundamental equation of mechanics; reference to co-ordinate axes...	67
71.	Application to a rigid solid...	69
72.	Application to a free rigid solid.....................................	71
73.	Interpretation of equations of translation and rotation.................	72
74.	Conditions in cases of constraint....................................	73
75.	Conditions of equilibrium...	74
76.	Object of analytical mechanics	74
77.	Equations T and R referred to centre of mass.......................	74
78.	Equations of translation; conclusions................................	75
79.	Illustration of conclusions..	76
80.	Equations of rotation.. ..	77
81.	Equations of rotation of a body having a fixed point; a fixed right line..	78
82.	Independence of the motion of translation and rotation................	79

General Theorem of Energy applied to a Free Rigid Solid.

Motion of Translation.

ART.		PAGE
83.	Translation under incessant forces.	79
84.	Velocity varies with co-ordinates of centre of mass.	80
85.	Rotation under incessant forces.	81
86.	Kinetic energy of translation and rotation.	82
87.	Translation under impulsion.	83
88.	Motion of translation: direct and inverse problem	83
89.	The direct problem.	83
90.	The inverse problem.	85
91.	Examples of the direct problem: (1) constant forces.	86
92.	(2) Motion due to gravity.	87
93.	(3) The trajectory in vacuo.	89
94.	(4) The trajectory in air.	94

Motion of Rotation.

95.	Moments of inertia.	97
96.	Radius and centre of gyration.	98
97.	The momental ellipsoid.	98
98.	Principal axes.	100
99.	Moment of inertia with reference to a parallel axis through the centre of mass.	101
100.	Discussion of the momental ellipsoids of a body.	102
101.	Determination of the moments of inertia.	104
	A uniform straight rod; circular arc; rectangular plate; triangle, elliptical area; ellipsoid; rectangular parallelopipedon.	
102.	Moments of inertia tabulated.	112

Instantaneous Axis.

103.	Equations of instantaneous axis.	113
104.	Direction of instantaneous axis.	114

Rotation of a Rigid Solid under Impulsion.

105.	Resultant axis and plane.	115
106.	Angular velocity about a principal axis.	116
107.	Angular velocity about the instantaneous and invariable axes.	116
108.	Position of the invariable plane.	117

ART		PAGE
109.	The rolling cone	119
110.	Discussion of the rolling cone	120
111.	The polhode and herpolhode	122
112.	Permanent axes of rotation	124
113.	Stability of rotation	124

Rotation of a Rigid Solid under Incessant Forces.

114.	Euler's equations of rotation	125
115.	Auxiliary angles	127
116.	Nutation and precessional motion	127
117.	The gyroscope	128
118.	Differential equations of motion of the gyroscope	130
119.	Nutation and precession of gyroscope	132

Impact.

120.	Definition of impact: compression and restitution	134
121.	Direct and central impact	135
122.	Oblique impact	137
123.	Impact against a fixed obstacle	137

Axis of Spontaneous Rotation.

124.	Equations of spontaneous axis	138
125.	Conditions for development of spontaneous axis	140
126.	Discussion of the spontaneous axis	140
127.	Axis and centre of percussion	141
128.	Reciprocity of the centres of percussion and spontaneous rotation for parallel impacts	142

Constrained Motion.

129.	Constrained motion defined	142
130.	Equations of constraint	142
131.	The normal reaction	144
132.	The theorem of energy applied as in free motion	146
133.	The value of the normal reaction	147
134.	Discussion for concave and convex curves	149
135.	Centrifugal force	150
136.	Change in apparent weight due to earth's rotation	152

ART.		PAGE
137.	Problems in constrained motion	152
138.	On an inclined plane	153
139.	Mechanical property of circular arcs	155
140.	On the arc of a cycloid	156
141.	On the arc of a circle	159

Constrained Motion about a Fixed Axis.

142.	The compound pendulum	160
143.	The equivalent simple pendulum	161
144.	Reciprocity of the axes of oscillation and suspension	162
145.	Minimum time of oscillation	162
146.	The simple seconds pendulum; Kater's method of finding its length	162
147.	Method by the reversible pendulum	165
148.	The value of the acceleration due to gravity	165
149.	Length of the equivalent simple pendulum	166
150.	The British standard of length	167
151.	Moment of inertia by the compound pendulum	167
152.	The conical pendulum	169
153.	The motion in azimuth of the conical pendulum	170

Equilibrium.

154.	Constraint necessary for a body in equilibrium	171
155.	The three cases of equilibrium; the first case	171
156.	The second case	172
157.	The third case	173
158.	Case of a free body illustrated; examples of constraint—(1) on a spherical concave surface; (2) the place of rest of a particle on a curve	173

The Potential.

159.	Attractions governed by the law of gravitation	178
160.	Component attractions	178
161.	The potential defined and explained	180
162.	Equi-potential surfaces	181
163.	The potential expressed in rectangular and polar co-ordinates	183
164.	Examples in attractions: a straight rod; circular arc; circular ring; circular plate; thin spherical shell; thick spherical shell	184
165.	The theorem of Laplace	191
166.	Poisson's extension of Laplace's theorem	192
167.	Another expression of the same theorem	193

Motion of a System of Bodies.

ART.		PAGE
168.	Translation of the centre of mass of the system..................	195
169.	Translation of its centre of mass with reference to a fixed origin; rotation of its centre of mass...	196
170.	Conservation of the motion of the centre of mass.................	197
171.	Reference to the solar system........................	198
172.	Conservation of moments; invariable axis and plane..............	199
173.	Conservation of areas..	200
174.	Relative acceleration..	201
175.	Differential equations of the relative orbit....	202

Central Forces.

176.	Central force defined...	203
177.	Laws of central forces..	203
178.	Differential equation of the orbit....	207
179.	Direct and inverse problem in central forces.....................	208
180.	Particular cases of the direct problem; central force attractive and varying directly as the distance.................	209
181.	Central force repellent...	212
182.	Central force attractive and varying inversely as the square of the distance..	212
183.	The velocity from infinity at any distance R.....................	215
184.	The velocity at any point of the orbit..........................	216
185.	The time of description of any portion of the orbit; the elliptical orbit; the parabolic orbit; the hyperbolic orbit..........................	216
186.	The anomalies..	222
187.	Illustration of the anomalies...................................	224

The Solar System

188.	The solar system defined..	225
189.	Kepler's laws.....	226
190.	Consequences of Kepler's laws...................................	226
191.	Law of universal gravitation....................................	228
192.	Planetary orbits..	230

THEORY OF MACHINES.

Resistances.

ART.		PAGE
193.	Resistances in machines	234
194.	Friction	234
195.	Coefficient of friction	235
196.	Problems involving friction	237
	(1) Motion on a plane surface	237
	(2) Friction on a trunnion	238
	(3) Friction on a circular pivot	240
	(4) Friction on a ring pivot	241
	(5) Friction of a cord on a cylinder	242
197.	Stiffness of cordage	244

Machines.

198.	Machine defined	246
199.	Theory of machines	247
200.	Use of fly-wheel	249
201.	Efficiency	250

Simple Machines.

202.	Gain and loss of power; equation of equilibrium	251
203.	The lever	253
204.	The common balance	255
205.	The wheel and axle	257
206.	The differential wheel and axle	260
207.	The pulleys; the fixed pulley	260
208.	The movable pulley	263
209.	The block and fall	264
210.	Other combinations of fixed and movable pulleys	266
211.	The inclined plane	266
212.	The work done in moving a body up a plane	268
213.	The wedge	268
214.	The screw	271
215.	The modulus of the screw	273
216.	The cord	275
217.	The differential equations of the funicular curve	276

ART.		PAGE
218.	The direction of the resultants of equilibrium forces acting on a cord.	277
219.	The ratio of the Intensities of these resultants...	278
220.	Application to a funicular curve.....................................	278
221.	The catenary curve...	279
222.	The common catenary....	279
223.	The directrix of the catenary.	281

MECHANICS OF FLUIDS.

Introductory.

224.	Classification of fluids..	283
225.	Perfect fluids defined...	283
226.	Definitions of stress, compression, compressibility and elasticity.....	283
227.	Pascal's principle.. ...	284

Laws of Perfect Gases.

228.	Action of heat on gases.......	284
229.	Fundamental equation of mechanical theory of heat................	285
230.	Laws of the gaseous state..	286
	(1) Boyle's or Mariotte's...	286
	(2) Charle's or Gay Lussac's.....................................	287
231.	Boyle's and Charles's laws combined....	288
232.	Absolute temperatures.........	288
233.	Differential equations of the specific heats........................	289
234.	Poisson's laws..	291
235.	Elasticity of volume under Boyle's law...............................	293
236.	Properties of actual gases..	293

Hydrostatics.

237.	Transmitted pressure on surface.......................................	294
238.	Pressure due to weight...	294
239.	Pressure on plane surface...	295
240.	Hydrostatic paradox..	296
241.	Centre of pressure...	297
242.	Graphical construction of centre of pressure..........	298
243.	Buoyant effort of fluids...	299
244.	Position of rest of submerged body...................................	300
245.	Specific gravity...	301

ART.		PAGE
246.	Stability of floating bodies	302
247.	Condition of stable equilibrium	304
248.	Metacentre	305

Hydrodynamics.

249.	Assumptions for theoretical discussion	305
250.	Euler's equations of fluid motion	305
251.	Equation of continuity	308
252.	Solution when the potential and velocity functions are exact differentials	310

Flow of Perfect Liquids through Vessels.

253.	General formula for velocity of discharge	311
254.	Two cases: (1) Upper surface at constant level	313
255.	(2) Vessel emptying itself	316
256.	Discharge through small orifice	317

Steady Flow of Fluids.

257.	Two cases; liquids and gases	318
258.	Bernouilli's law	319
259.	Uniform flow of liquids	320
260.	Steady flow of gases	321

Equilibrium of Liquids in Motion.

261.	General formula for surface of uniform pressure	323
	Ex. 1. Liquid rotating about a vertical axis	324
	Ex. 2. Equilibrium surface of earth	326

Hydraulics.

262.	Viscosity	328
263.	Experimental co-efficients in the flow of water	329
264.	Mouthpiece of Borda	331
265.	Cylindrical mouthpiece	332
266.	Divergent mouthpiece	335
267.	The hydraulic gradient	336

Applications.

268.	Hydraulic machines	338

Water Motors.

269.	Water-wheels	339
270.	Turbines	341
271.	Theory of turbines	342

Pumps.

ART.		PAGE
272.	The sucking pump	344
273.	Play of the piston	345
274.	The lifting pump	347
275.	The force-pump	347
276.	Quantity of energy expended in pumping	348
277.	The centrifugal pump	351
278.	Double-acting force-pump	351
279.	The hydraulic ram	351
280.	The siphon	352
281.	The air-pump	353
282.	The mercurial air-pump	355
283.	The atmosphere and its pressure	356
284.	The barometer	357
285.	Standard atmosphere	358
286.	Density of air	359
287.	Height of homogeneous atmosphere	359
288.	The barometric formula	360
289.	Conditions to be observed in its use	362

APPENDIX.

Table I.	Densities and specific gravities of substances	363
Table II.	The metric system	367
Table III.	To convert metric into U. S. measures, and conversely	368
Table IV.	Gravity—the values of g and L	370
Table V.	Friction	371
Table VI.	Stiffness of cordage for white and tarred rope	372
Table VII.	Relative density of water at different temperatures	373

The Greek Alphabet is here inserted to aid those who are not already familiar with it, in reading the parts of the text in which its letters occur.

Letters.	Names.	Letters.	Names.
$A\ \alpha$	Alpha	$N\ \nu$	Nu
$B\ \beta$	Bēta	$\Xi\ \xi$	Xi
$\Gamma\ \gamma$	Gamma	$O\ o$	Omicron
$\Delta\ \delta$	Delta	$\Pi\ \pi$	Pi
$E\ \epsilon$	Epsilon	$P\ \rho$	Rho
$Z\ \zeta$	Zēta	$\Sigma\ \sigma\ \varsigma$	Sigma
$H\ \eta$	Eta	$T\ \tau$	Tau
$\Theta\ \theta\ \vartheta$	Thēta	$\Upsilon\ \upsilon$	Upsilon
$I\ \iota$	Iōta	$\Phi\ \phi$	Phi
$K\ \kappa$	Kappa	$X\ \chi$	Chi
$\Lambda\ \lambda$	Lambda	$\Psi\ \psi$	Psi
$M\ \mu$	Mu	$\Omega\ \omega$	Omega

MECHANICS OF SOLIDS.

1. *Mechanics* treats of the equilibrium and motion of bodies, or their elements, under the action of forces.

2. The most ordinary observation shows that changes are constantly taking place in matter. These changes are assumed to be due to the action of force, and a complete analysis of the various phenomena would make known the particular force or forces at work. To make the analysis we must cultivate the faculty of observation and acquire skill in experimentation. But as no one can possibly repeat all the experiments, nor observe all the phenomena which at present form the data upon which Mechanics is based, we must accept the certified facts and the conclusions derived therefrom, until we are sufficiently instructed in the science to form for ourselves a rational judgment as to their truth. In this we exercise a proper faith in the honesty, accuracy and ability of those who have devoted their lives to the study of the natural sciences. It is well to remember that the accepted laws which are assumed to govern the changes in the state or condition of matter can never be exactly verified by experiment, because of the inaccuracy of the experimenter and the imperfections of the appliances by which the results are measured. But whenever a stated law appears to conform more nearly to the observed results as the experimenter becomes more skilful and the apparatus more perfect, we accept it as the governing and limiting law for this class of phenomena.

Matter.

3. Of the ultimate nature of matter we are ignorant; but from close observation of natural laws it has been assumed:

(1) That every material substance is composed of one or more *simple substances* or *elements*, so called because they have thus far resisted simplification by subdivision;

(2) That each of these simple substances is composed of very minute, but finite and definite, portions, called *atoms ;*

(3) That in any substance, simple or compound, two or more of these atoms are, in general, so united as to form the smallest portion that can exist by itself and remain the same substance. This combination of atoms is called a *molecule.*

4. To illustrate: hydrogen and oxygen are simple substances, and two atoms of hydrogen unite with one of oxygen to form a molecule of water, which is a compound substance. No quantity of water less than the molecule can exist, but if the molecule be divided it becomes hydrogen and oxygen. In Mechanics the molecule is therefore considered as the elementary mass.

5. *Mass* is a term used to express the quantity of matter in a body, and its numerical value will depend upon the quantity of matter assumed as the unit mass.

6. *Density* is that property of a body by which the quantity of matter in the unit volume is determined. It varies in different bodies according to the nearness and mass of their constituent molecules. When a body is compressed its density is increased because its molecules are brought nearer together. There are therefore more of them, and hence more mass, in the unit volume than before. The contrary is the case when the body is expanded. The density of a body is measured by the number of mass units in its unit volume. In comparing densities, however, the standard is that of pure water under established conditions.

7. The mass of a body is therefore directly proportional *to* the product of its volume and its density. If the unit mass be assumed to be the mass of a unit volume of matter at unit density, then the mass of any homogeneous body will be given by the equation

$$M = V\delta, \quad \ldots \ldots \ldots \quad (1)$$

in which M is the mass in mass units, V the volume in volume units, and δ the density of the body.

8. *Inertia* is defined to be that property of a body by which it continues in its particular state of rest or rectilinear uniform motion until the action of some force produces a change in one or both of these states.

Force.

9. *Force* is that which produces, or tends to produce, any change in the state of matter with respect to rest or motion.

The *intensity* of a force is its capacity to produce pressure. The *point of application* is the molecule of the body to which the force may be considered as directly applied. The *action-line* is the right line which the point of application would describe if it were free to move from rest under the action of the force alone.

A force is said to be completely given when its intensity, action-line and point of application are known; for if any one of these be varied the resulting effect of the force will in general be changed.

10. *Molecular Forces.*—Every molecule of a body is assumed to be the locus of a force which is sometimes attractive and sometimes repellent according to the particular circumstances of its development. By virtue of this force the molecule unites itself with its adjacent molecules, or tends to separate itself from them; and by means of these forces the mass assumes either the solid, liquid or gaseous state, under particular conditions of temperature and external pressure. When an extraneous force is applied to a body these molecular forces manifest themselves and oppose its action. After the extraneous force is withdrawn

the molecular forces come again into equilibrio, and the molecules either resume their primitive positions of equilibrium or assume new ones. For example, when a solid bar is subjected to a stress of elongation the molecules of consecutive cross-sections will increase their distances from each other, and during this state of separation the molecular forces will be attractive; but when there is a stress of compression the corresponding molecules are brought nearer to each other, and during this state their forces are repellent. In either case the aggregate intensity of the molecular forces is equal to that of the applied stress. When the stress is withdrawn, the molecular forces are balanced and the molecules return to their primitive positions, provided they have not been forced beyond their elastic limits.

11. *Elasticity* is that property of a body by virtue of which the molecular forces restore, or tend to restore, the molecules to their primitive relative positions when they have been moved from these positions by the action of some force.

The elasticity is said to be perfect when the body always requires the same force to keep it at rest in the same volume, shape and temperature, through whatever variations of volume, shape and temperature it may have passed.

Every body has some degree of elasticity of volume. All fluids possess great elasticity of volume. If a body possess any degree of elasticity of shape it is called a solid; if none, a fluid. While the elasticity of shape is very great for many solids, it is not perfect for any. The degree of distortion within which elasticity of shape is sensibly perfect is limited in every solid; when the distortion passes beyond this limit the body either breaks or receives a permanent *set;* that is, such a molecular displacement that it does not return to its original figure when the distorting force is removed.

12. *Gravitation.*—It is assumed that any body in the universe attracts any other body with an intensity which varies directly as the product of their masses and inversely as the square of the distance which separates them; also that this attraction is mutual, or that the intensity of the attraction of a body a for a body b is exactly equal and directly contrary to that of *b* for *a*. Let *m* and *m'* be the number of mass units in the bodies *a* and *b*, Fig. 1; *r* the distance between their centres, and μ the attraction of one mass unit for another mass unit at a unit's distance apart; then the intensity of the mutual attraction of the two bodies is given by

FIG. 1.

$$G = \frac{mm'}{r^2}\mu; \quad \ldots \quad \ldots \quad (2)$$

for, each of the mass units of *m* attracts each of those of *m'* with an intensity μ at the distance unity; and as this intensity varies inversely as the square of the distance, with an intensity $\frac{\mu}{r^2}$ at the distance *r*; therefore the *m* units of one body will attract the *m'* units of the other with an intensity $\frac{mm'}{r^2}\mu$, and conversely the *m'* units will attract the *m* units with an equal intensity. The mutual attraction existing between any two bodies is *a single force* whose stress or motive effect on each body can be determined separately.

Motion.

13. *Kinematics* is that branch of pure mathematics in which the properties of motion are considered without reference to its cause. Motion is the state of a body when it is changing its place with respect to an origin. A body is said to be at rest with respect to an origin, or at relative rest, when it remains at the same distance and in the same direction from the origin. Considering the diurnal and annual motion of the earth, that of the solar system through space, and the proper motion of the fixed stars, we see that a state of absolute rest is unknown for any body in the universe. Rest is therefore wholly relative.

14. Motion is continuous; for a body cannot pass from one position to another without occupying all intermediate positions in the path described. Motion may be uniform or varied. It is uniform when the moving body describes equal spaces in equal successive portions of time, no matter how small these intervals of time may be. When this condition is not fulfilled the motion is varied.

15. *Velocity* is the rate of motion. Its *measure*, at any instant, is the distance that would be described in the next subsequent unit of time, were the motion to continue unchanged during that unit. Hence the laws of uniform motion are embodied in the equation

$$s = vt, \quad \ldots \ldots \ldots \ldots (3)$$

s being the distance described in the time t, v the constant velocity, and t the units of time since the epoch $t = 0$. The principles of the calculus and the definition of velocity give, for the measure of uniform or varied velocity at any instant,

$$v = ds \times \frac{1}{dt} = \frac{ds}{dt}. \quad \ldots \ldots \ldots (4)$$

Velocity is therefore measured by a distance along the direction of the motion at the instant considered, and hence it may be graphically represented by a right line. The projections of this right line on co-ordinate axes represent the component velocities in these directions. If $\frac{ds}{dt}$ be the measure of the velocity at any instant, then the component velocities along the axes will be measured by $\frac{dx}{dt}, \frac{dy}{dt}, \frac{dz}{dt}$, and when the axes are rectangular, Fig. 2, their relations to each other are given by,

$$\frac{ds}{dt} = \sqrt{\frac{dx^2}{dt^2} + \frac{dy^2}{dt^2} + \frac{dz^2}{dt^2}}. \quad . \quad (5)$$

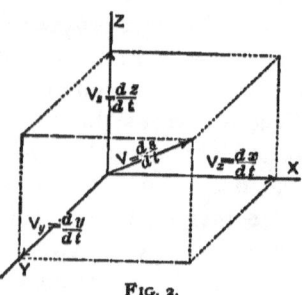

FIG. 2.

This is known as the *principle of the parallelopipedon of velocities*, and may be thus stated: *If the three edges of a parallelopipedon which meet at one vertex represent component velocities, the diagonal through this vertex represents the resultant velocity.* It is employed in the resolution and composition of velocities.

16. *Acceleration* is the rate of change of velocity. Whether constant or variable, it is measured by the increment of velocity in a unit of time, supposing the acceleration to remain constant for that unit, and the same as at the instant considered. Hence its measure is

$$\alpha = \frac{dv}{dt} = \frac{d^2s}{dt^2}. \quad \ldots \quad \ldots \quad (6)$$

When the velocity is increasing, the acceleration is regarded as positive; and when decreasing, as negative. Acceleration may be graphically represented by a right line, since it is measured by a velocity, and the projections of its rectilinear representative on co-ordinate axes will represent its component accelerations in these directions. Hence we have $\frac{d^2x}{dt^2}$, $\frac{d^2y}{dt^2}$, $\frac{d^2z}{dt^2}$, as the component accelerations of $\frac{d^2s}{dt^2}$ when the latter is the acceleration along the right-line path s. If the path be a plane curve, and ρ the radius of curvature at any point, we have from the calculus

$$(d^2x)^2 + (d^2y)^2 + (d^2z)^2 - (d^2s)^2 = \frac{ds^4}{\rho^2}. \quad \ldots \quad (7)$$

Dividing both members by dt^4 and reducing, we have

$$\sqrt{\left(\frac{d^2x}{dt^2}\right)^2 + \left(\frac{d^2y}{dt^2}\right)^2 + \left(\frac{d^2z}{dt^2}\right)^2} = \sqrt{\left(\frac{d^2s}{dt^2}\right)^2 + \frac{V^4}{\rho^2}} \quad \ldots \quad (8)$$

The first member is the resultant acceleration, and its value in the second member is compounded of two accelerations at right angles to each other; the one, $\frac{d^2s}{dt^2}$, in the direction of the path at the instant considered, and the other therefore in the direction of the radius of curvature towards the centre. When the path s becomes a right line ρ becomes infinite, and the second component acceleration zero.

17. *Angular velocity* is the rate of motion about a centre, and *angular acceleration* is the rate of change of the angular velocity. Representing the first by ω, and the second by $\dfrac{d\omega}{dt}$, their measures are given by

$$\omega = \frac{d\theta}{dt}, \quad . \quad . \quad (9) \qquad \frac{d\omega}{dt} = \frac{d^2\theta}{dt^2} = \omega\frac{d\omega}{d\theta}, \quad . \quad . \quad (10)$$

in which θ is the angle made by the radius vector with the line of reference through the centre. The tangential linear velocity of a point at a unit's distance from the axis is therefore a measure of the angular velocity.

The unit angular velocity is that of a point describing the unit angle ($57°.29578 +$, called the radian) uniformly in a unit of time.

Physical Units.

18. *The British System.*—The British *unit of length* is the *foot*. It was first established as a standard by taking it to be a certain fraction of the length of the simple seconds pendulum at London, but is now defined to be the third part of the distance between two marks on the gold plugs of the *Standard Yard* deposited in the Exchequer at London.

The unit of time is either the *sidereal* or *mean solar second*, both of which are determined from the uniform rotation of the earth on its axis. This unit is international. Unless otherwise stated the *mean solar second* is assumed as the unit.

The British *unit of mass* is a certain platinum cube called the *Pound*, deposited in the Exchequer at London, and made the standard unit of mass in Great Britain by act of Parliament. By means of its copies the masses of other bodies may be determined.

The units of space, time and mass are arbitrary units, and serve to determine the other units of the system, which are called *derived units*.

The Unit of Force.—Gauss has defined the *absolute unit of force* to be that force which, acting on a given unit of mass for a unit of time, would generate in it a unit of velocity. By this definition, the unit force can be derived from any standard units of mass, time and velocity with equal facility. The *British unit of force* is that force which, acting on the Pound mass for one second, would generate in it a velocity of one foot per second. Since the Pound is an invariable mass, this establishes an invariable British standard unit of force. This unit is called the *Poundal*.

19. *Other Units.*—The *unit of velocity* is a velocity of unit distance per unit time, and is, therefore, one foot per second. Similarly, the *unit of acceleration* is an acceleration of one foot per second. The *unit of area* is the square foot; that of *volume*, the cubic foot; and that of *density*, the density of the unit of matter occupying a cubic foot of volume.

It will be shown later that the weight of a given mass on the earth's surface varies with the latitude and the height above the sea-level. Therefore weight cannot be taken as an invariable standard of force; but as the variations in the weight of any mass are proportionally small, the weight of the British unit of mass is generally taken as the unit of force for ordinary purposes.

20. *The C. G. S. System.*—The *French* or *Centimeter, Gram, Second system*, is named from its length, mass and time units.

The *unit of length* in this system is the *centimeter*, derived from the meter which was formerly supposed to be equal to one ten-millionth of the quadrant of the Paris meridian line, but is now definitely fixed by a standard meter in the Archives at Paris.

The *unit of mass* is the *gram*, derived from the *kilogram* which was originally defined as the quantity of matter in a liter of pure water at the temperature of maximum density, but is now determined by existing standards.

The unit of force in this system, called the *Dyne*, is that force which, acting on the gram mass for one second, would generate a velocity of one centimeter per second.

Stresses and Motive Forces.

21. Nothing is known of the inherent nature of force; but the intensities of forces are assumed to be proportional to their effects under precisely similar circumstances of action, and can be estimated by comparing these effects. Forces are classed as *stresses* or *motive forces* according as their effects are *strains* or *changes of state with respect to rest or motion*. That branch of Mechanics which treats of stresses and their effects is called *Statics*, and that which treats of motive forces is called *Kinetics*.

22. *Measure of the Intensity of a Stress.*—When a solid bar is subjected to a stress within its elastic limit, experiment shows that the elongations are directly proportional to the *intensity* of the elongating stress, and to the *original length* of the bar; and inversely proportional to the *area of cross-section*, supposed constant throughout the experiment, and to a *coefficient depending on the material of the bar*. These experimental laws are expressed by

$$\lambda = \frac{Il}{Es}; \qquad \qquad (11)$$

in which λ is the elongation due to the stress whose intensity is I, l the original length of the bar, s the constant area of cross-section, and E a coefficient depending on the material of the bar.

If this law be supposed to hold good for all longitudinal stresses until $\lambda = l$, we have

$$I = E \qquad \qquad (12)$$

for a bar of unit area of cross-section. E is therefore the intensity of that stress which, applied in the direction of the length, would elongate a bar of unit area of cross-section to double its original length, or compress it to zero length under the assumed law. Making $l = 1$, and $s = 1$, in Eq. (11), we have

$$E = \frac{I}{\lambda}; \qquad \ldots \ldots \ldots \quad (13)$$

whence E may also be defined to be the ratio of the stress to the elongation produced by it in a bar of unit length and cross-section. It is called Young's Modulus or Coefficient of Longitudinal Elasticity. From Eq. (13) we have

$$I = E\lambda. \qquad \ldots \ldots \ldots \quad (14)$$

Hence, within elastic limits, the intensities of stresses are assumed to be directly proportional to the elongations or compressions which they would produce in a given bar, when applied longitudinally. This is the principle of the common spring-balance.

23. *Measure of the Intensity of a Motive Force.*—Let P be the type symbol of a *force*, and I that of its *intensity*. Experiment and observation show conclusively that if the force act upon different free masses m, m', m'', etc., I being constant, then will

$$m : m' :: \frac{d^2s'}{dt^2} : \frac{d^2s}{dt^2}; \qquad \ldots \ldots \quad (15)$$

that is, *when a constant force acts upon different free bodies, the accelerations are inversely as their masses.*

Also that if two forces of different intensity act upon the same free mass, we will have

$$I : I_{\prime} :: \frac{d^2s}{dt^2} : \frac{d^2s_{\prime}}{dt^2}; \qquad \ldots \ldots \quad (16)$$

or, *the accelerations are directly as the intensities of the forces.*

From the first of these principles we get, for forces of equal intensities,

$$m\frac{d^2s}{dt^2} = m'\frac{d^2s'}{dt^2} = m''\frac{d^2s''}{dt^2} = \text{etc.}; \quad \ldots \quad (17)$$

and from the second, for forces of different intensities,

$$I : I_{,} :: m\frac{d^2s}{dt^2} : m\frac{d^2s_{,}}{dt^2}, \quad \ldots \ldots \quad (18)$$

or

$$I : I_{,} :: m\frac{d^2s}{dt^2} : m'\frac{d^2s_{,}'}{dt^2}. \quad \ldots \ldots \quad (19)$$

From Eq. (17) we see that *for any constant force the product of the mass and acceleration of the free body on which it acts is a constant quantity;* and, from Eq. (19), that *for different constant forces the intensities are proportional to such products.* Hence $m\dfrac{d^2s}{dt^2}$ fulfils all the requirements of a measure, and we may write

$$I = m\frac{d^2s}{dt^2}. \quad \ldots \ldots \ldots \quad (20)$$

This evidently measures the intensity of *any motive force;* for, if the force be variable, the acceleration at any instant is the change in the velocity which would take place in a unit of time, provided the force were to remain constant during that unit. Hence *the intensity of a motive force is measured by the product of the mass of the free body upon which it acts by the acceleration due to the force.* The product of a mass by its velocity is called its "quantity of motion," "quantity of velocity," and "momentum" by different authors. The term *momentum* is adopted in this text.

Since

$$m\frac{d^2s}{dt^2} = m\frac{dv}{dt} = \frac{d(mv)}{dt}, \quad \ldots \ldots \quad (21)$$

we see that the measure of the intensity of a motive force is *the rate of change of the momentum taken with respect to the time.*

24. *Impulsive Forces.*—A force which acts on a body for a very short time, as in the case of a blow, is called an *impulsive force* or *impulsion,* while one whose action is continuous is called an *incessant force.* If an impulsive force were to be measured as in the case of incessant forces, dv would be great compared with dt, and the expression $\frac{dv}{dt}$ would generally be a velocity inconveniently large. It would also, in general, be impracticable to measure the duration of the action of such a force. Hence the measure of the intensity of the impulsion is assumed to be *the momentum generated during the whole time of action of the force*, no matter how long or short this time may be, and *not that which would be generated in a unit of time*. Therefore if I_t be the intensity of an impulsion, we have for its measure

$$I_t = M\frac{ds}{dt} = Mv = M(v_1 - v_0), \quad \ldots \quad (22)$$

in which v_0 is the velocity of M at the instant the impulsion began its action on the body, and v_1 the velocity when its action is completed. It is evident that this is also the measure of the intensity of an incessant force which would, in a unit of time, generate the same momentum as that which is actually produced by the impulsive force. There is therefore no distinction between incessant and impulsive forces save that relating solely to the duration of their action.

25. *Action of Forces upon Free Bodies.*—*A free rigid solid* is a body perfectly free to move under the action of any extraneous force whatever, its molecules being so connected that no change of relative position is possible among them. No such body occurs in nature. All bodies change their form, either temporarily or permanently, when subjected to the action of extraneous forces; and the results deduced in Mechanics, under the supposition that bodies are free rigid solids, are not in strict accord with those observed in actual masses.

If an isolated molecule could receive the action of an extraneous force without the counterbalancing influences of other molecules, it would immediately begin to acquire accelerated motion, and continue to do so during the action of the force; after which it would move with uniform motion, until again subjected to the action of force. This is the simple consequence of the assumed definition of force. Were the body acted on the hypothetical free rigid solid, the increments of velocity of the different molecules would be simultaneous.

Let the body be a free solid, but not rigid, and suppose that m, m', etc., Fig. 3, represent a file of its molecules along any direction, their positions being fixed by the molecular attractions which appertain to the body. Let the force P act on m to move it towards m'. As soon as m approaches m' the molecular forces on m' will be unbalanced, and m' will be moved towards the next molecule in order, seeking a new position of equilibrium nearer to m''. In like manner each molecule will in succession take up its change of position and of state, until the last molecule of the file, m'', is reached. Some interval of time is therefore required before the full effect of the action of the force, as exhibited in the motion of the body as a whole, is manifested.

So long as the force continues to act the molecules are in a state of strain, being nearer each other than they were before; and the distance between each molecule and the one on its left is less than that between it and the one on its right. The difference between the molecular forces corresponding to these distances is the force which is employed in giving the molecule its acceleration, and its intensity is measured by $m\frac{d^2s}{dt^2}$.

Thus we see that the molecular forces serve to distribute the effect of the extraneous force throughout the body. Because of this action *the point of application of a force may be taken anywhere on its line of direction within the limits of the body.*

26. *Representation of Forces.*—Since a force is completely given when its intensity, direction and point of application are known (Art. 9), we may graphically represent a force by a portion of its action-line equal in length to the number of units in the intensity of the force. One end of this portion is taken at the point of application, and the direction of the action is indicated by an arrow-head, placed generally at the other end. Thus the right line in Fig. 4 represents a force, since when the line is given the force is given. The scale according to which the line is constructed is indicated in the figure.

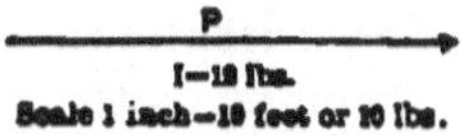

P
1—12 lbs.
Scale 1 inch = 10 feet or 10 lbs.

Fig. 4.

It is sometimes convenient, in order to avoid confusing a figure, to take the extremity at the arrow-head as the point of application; but when such departure from the general rule is made, the change is evident from the construction of the figure.

27. *Rectangular Components of a Force.*—Let there be three

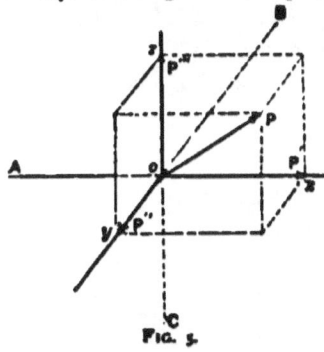

Fig. 5.

perpendicular cords, AO, BO and CO, joined at O, Fig. 5, and let their directions be taken as a set of co-ordinate axes. The tensions on the cords, caused by the force P, will be given by the projections Ox, Oy, Oz, of the intensity of P on the axes. This follows from the assumption of the right line as the representative of a force, and its truth has been conclusively shown by experiment. The forces P', P'' and P''' are called *rectangular components* of the force P.

Let α, β, γ be the type-symbols of the angles made by the action-line of a force with the co-ordinate axes x, y, z respectively, and let X', Y', Z' represent the component intensities of I in the directions of these axes. Then we have

$$\left. \begin{array}{l} X' = I \cos \alpha; \\ Y' = I \cos \beta; \\ Z' = I \cos \gamma. \end{array} \right\} \quad \ldots \ldots \quad (23)$$

28. *Resultant of a System of Forces.* The *resultant of a system of forces* is a single force which will produce the same effect as all the forces of the system acting together. The forces of the system are called *components* of the resultant.

It is obvious that while a system of forces may have but a single resultant, the resultant may have any number of systems of components.

Let R be the intensity of the resultant, and a, b, c the angles which its action-line makes with the co-ordinate axes respectively. Since the effect of the resultant is the same as that of the system of forces, *its component in any direction must be equal to the sum of the components of the forces of the system in that direction,* and we have

$$\begin{aligned} R \cos a &= I \cos \alpha + I' \cos \alpha' + \text{etc.} = \Sigma X' = X; \\ R \cos b &= I \cos \beta + I' \cos \beta' + \text{etc.} = \Sigma Y' = Y; \\ R \cos c &= I \cos \gamma + I' \cos \gamma' + \text{etc.} = \Sigma Z' = Z. \end{aligned} \quad (24)$$

Squaring and adding, we have

$$R^2 (\cos^2 a + \cos^2 b + \cos^2 c) = X^2 + Y^2 + Z^2, \quad . \quad . \quad (25)$$

or

$$R = \sqrt{X^2 + Y^2 + Z^2}. \quad . \quad . \quad . \quad . \quad . \quad (26)$$

From Eqs. (24) we have

$$\cos a = \frac{X}{R}; \quad \cos b = \frac{Y}{R}; \quad \cos c = \frac{Z}{R}. \quad . \quad . \quad (27)$$

The equations of a right line making the angles a, b, c with the axes are

$$\frac{x - x'}{\cos a} = \frac{y - y'}{\cos b} = \frac{z - z'}{\cos c}. \quad . \quad . \quad . \quad . \quad (28)$$

If the co-ordinates of the point of application of the resultant be x', y', z', the equations of its action-line become

$$\frac{R}{X}(x - x') = \frac{R}{Y}(y - y') = \frac{R}{Z}(z - z'), \quad . \quad . \quad (29)$$

or

$$\frac{x - x'}{X} = \frac{y - y'}{Y} = \frac{z - z'}{Z}. \quad . \quad . \quad . \quad . \quad (30)$$

29. *The Parallelogram of Forces.*—Let the system be composed of two forces with a common point of application. Take this point as the origin, Fig. 6, and let the forces P' and P'' lie in the plane XY. The rectangular components of P' are ox' and oy', and those of P'' are ox'' and oy'', neither having any component perpendicular to their plane. The rectangular components of the resultant are

$$ox' + ox'' = ox; \quad \ldots \quad (31)$$
$$oy' + oy'' = oy; \quad \ldots \quad (32)$$

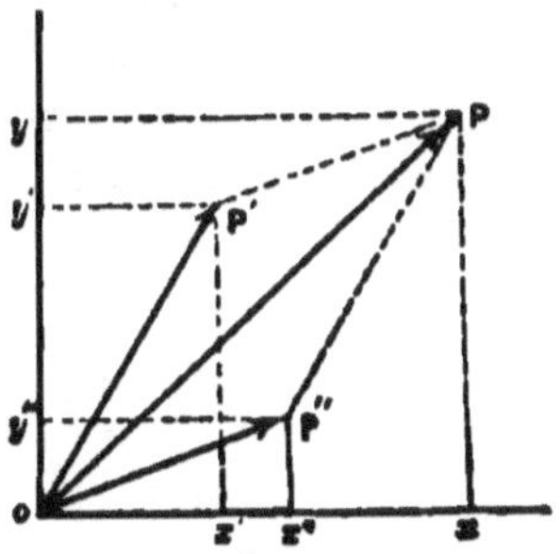

Fig. 6.

and the resultant of P' and P'' is the resultant of ox and oy, which is P. But we see from the construction that P is the diagonal of a parallelogram constructed on P' and P'', and that it passes through their common point.

We therefore conclude *that if two forces have a common point of application, and a parallelogram be constructed on their linear representatives, their resultant is completely represented by that diagonal of the parallelogram which passes through their point of application.*

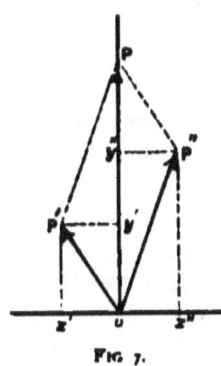

Fig. 7.

Let P be taken to coincide with the axis of Y, as in Fig. 7. The components of P' and P'' in the direction of X are equal to each other in intensity, and as they act in opposite directions they counteract each other. The sum of the intensities of the components in the direction of Y is represented by the length of the diagonal of the parallelogram constructed on P' and P'', and we have the resultant of the system represented by that diagonal, as before.

30. Since any side of a triangle is the diagonal of a parallelogram constructed on the other two sides, we conclude that if two sides of a triangle represent the *intensities* and *directions* of two component forces, the other side will represent the *intensity* and *direction* of their resultant. We assume, therefore, that all the trigonometric relations existing between the angles and sides of a triangle are true of the directions and intensities of a resultant and its pair of components.

These relations are

$$R^2 = I'^2 + I''^2 + 2I'I'' \cos(\phi' + \phi''); \quad \ldots \quad (33)$$

$$\left. \begin{array}{l} I' = \dfrac{R \sin \phi''}{\sin \delta}; \\[2mm] I'' = \dfrac{R \sin \phi'}{\sin \delta}; \end{array} \right\} \quad \ldots \ldots \ldots (34)$$

$$I' : I'' :: \sin \phi'' : \sin \phi'; \quad \ldots \ldots (35)$$

$$\left. \begin{array}{l} \sin \tfrac{1}{2} \phi' = \sqrt{\dfrac{(S-I')(S-R)}{RI'}}; \\[2mm] \sin \tfrac{1}{2} \phi'' = \sqrt{\dfrac{(S-I'')(S-R)}{RI''}}; \\[2mm] \cos \tfrac{1}{2} \delta = \sqrt{\dfrac{(S-I')(S-I'')}{I'I''}}; \end{array} \right\} \quad \ldots \ldots (36)$$

in which ϕ', ϕ'' and δ are the angles which P' makes with R, P'' with R, and P' with P'', respectively, and $S = \dfrac{R + I' + I''}{2}$.

31. Any number of forces having a common point of application, or any number of forces lying in the same plane but having different points of application, may be combined by the parallelogram of forces. Thus, in the first case, find the resultant of any two, then combine this resultant with one of the other forces, and so on until all have been combined. The last resultant is the resultant of the system.

In the second case, prolong the action-lines of two of the

forces until they meet. Take their intersection as a common point of application, (Art. 25), and then proceed as above.

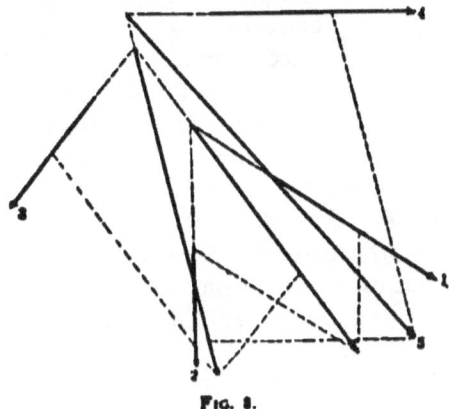

Fig. 8.

Fig. 8 is an illustration of such a combination, 5 being the resultant of 1, 2, 3 and 4.

32. *The Parallelopipedon of Forces.*—The principle of the parallelogram of forces is readily extended to that of the *parallelopipedon of forces*. Thus, in Fig. 9, P^{iv} is the resultant of P' and P'',

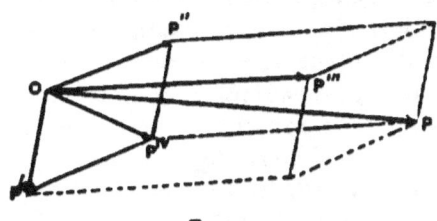

Fig. 9.

and P of P^{iv} and P'''. P is therefore the resultant of P', P'' and P'''. This principle is thus stated: *If three forces have a common point of application, and a parallelopipedon be constructed on their linear representatives, their resultant is completely represented by that diagonal of the parallelopipedon which passes through their point of application.*

33. *The Moment of a Force.*—If a body be free to rotate about a fixed point it is evident that any force whose action-line does not pass through the point will produce such rotation. Thus if o, Fig. 10, be a fixed point, the force P tends to rotate the body about o. This tendency is directly proportional to the intensity of the force and to the perpendicular distance from the action-line of the force to the point, and hence to their product, Ip. The product Ip is therefore the measure of the capacity of the force to produce rotation about o, and is called the moment of P with respect to o. Ip is also the moment of P with respect to an axis through o and perpendicular to the plane of o and the action-line of the force. This axis is called the moment axis of the force with respect to o. The point o is called the centre of the moment, and p the lever-arm of the force or of the moment.

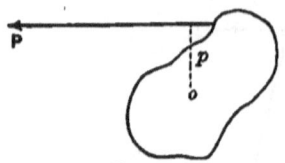

Fig. 10.

To find the moment of P with respect to any other axis through o, we make use of the principle that the moment of a force with respect to any axis is equal to the sum of the moments of its components. The moment of any component in the plane of the axis is zero. Hence if we resolve the force into two components, one of which is in a plane containing the axis and the other perpendicular to this plane, the moment of the force is equal to the moment of the perpendicular component. We therefore have the following rule for obtaining the moment of a force with respect to an axis: *Resolve the force into two components, one of which shall be perpendicular to the axis and the other in a plane containing the axis; multiply the intensity of the perpendicular component by the perpendicular distance between its action-line and the axis.* Any axis which is oblique to the action-line of the force is called a *component axis.*

It is generally most convenient to multiply the distance from the point of application to the axis, by the intensity of that component of the force which is normal to the plane containing the point of application and the axis.

STRESSES AND MOTIVE FORCES. 21

34. *Representation of Moments.*—We may write

$$Ip = I' \times 1 = I'. \quad \ldots \quad \ldots \quad (37)$$

That is, the moment measures *the intensity of a force* which would, with a lever-arm of unity, have the same effect as the given force to produce rotation about the axis. Therefore moments and forces may be measured by the same unit, and a definite portion of a right line may be taken to represent a moment. To indicate the centre of the moment and the axis, the moment representative is laid off from the centre of the moment on the axis. To show the direction of motion about the axis the line is terminated by an arrow-head, and is laid off in such a direction from the centre of the moment that the motion shall appear right-handed as one looks along the axis from the arrow-head toward the centre of the moment. Thus, in Fig. 11, the right line indicates the *intensity of the moment*, the *axis*, the *centre of the moment*, and the *direction of rotation*.

35. *The Moments of a Force with respect to Co-ordinate Axes.*—To find expressions for the moments of a force with respect to any set of rectangular coordinate axes, let P, Fig. 12, represent the force, and x', y', z' the co-ordinates of its point of application. $I \cos \gamma$ has no moment with respect to the axis of z, and the moment of P with respect to this axis is evidently equal to the sum of the moments of $I \cos \alpha$ and $I \cos \beta$. The lever-arm of $I \cos \alpha$ is y', and hence its moment is $I \cos \alpha \, y'$. The lever-arm of $I \cos \beta$ is x', and its moment is $I \cos \beta \, x'$. These two moments tend to produce rotation in opposite directions. Calling moments positive when

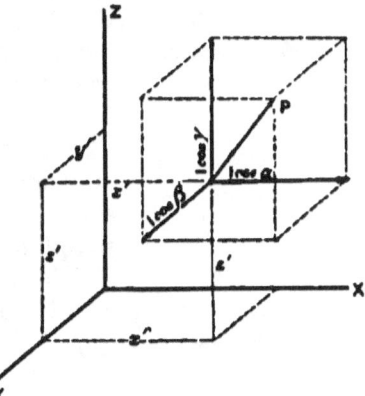

Fig. 12.

their representatives are laid off in positive directions along the axes, we have for the moment of P with respect to z

$$I(x' \cos \beta - y' \cos \alpha) \ldots \ldots \ldots (38)$$

Similarly, for the moments with respect to the axes of y and x, we have

$$\left. \begin{array}{l} I(z' \cos \alpha - x' \cos \gamma); \\ I(y' \cos \gamma - z' \cos \beta). \end{array} \right\} \ldots \ldots (39)$$

For a system of forces we have

$$\left. \begin{array}{l} R(x \cos b - y \cos a) = \Sigma I(x' \cos \beta - y' \cos \alpha) = L; \\ R(z \cos a - x \cos c) = \Sigma I(z' \cos \alpha - x' \cos \gamma) = M; \\ R(y \cos c - z \cos b) = \Sigma I(y' \cos \gamma - z' \cos \beta) = N; \end{array} \right\} (40)$$

in which R represents the intensity of the resultant, x, y and z the co-ordinates of its point of application, and a, b and c the angles which its action-line makes with the axes.

36. *Composition and Resolution of Moments.*—Let oM, Fig. 13, represent a moment, and let it be required to find its component with respect to any axis through the centre of moments, as oM'. We may assume that the moment oM is due to a force applied at o'' and perpendicular to the plane of the figure. This force will have the lever-arm $o'o''$ with respect to the axis oM', and its moment with respect to that axis is therefore, oo'' being represented by p,

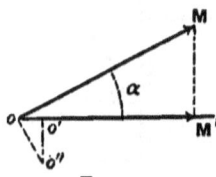

Fig. 13.

$$I(p \cos \alpha) = (Ip) \cos \alpha = oM \cos \alpha. \ldots (41)$$

But $oM \cos \alpha$ is the projection of oM on the axis oM'. Hence we see that the component of a moment, with respect to any axis through the centre of the moment, is represented by the projection of the moment on that axis. From the assumption of the rectilinear representative of a moment we see that the principles of the parallelogram and parallelopipedon of moments follow, as in the case of forces. Figs. 6, 7 and 9 may therefore be taken to illustrate the composition of moments, P, P', P'' and P''' representing moments with a common centre at o.

37. Assuming a common centre of moments for a system of forces, let Rp represent the resultant moment, R being the intensity of the resultant of the forces, and p its lever-arm; and let l, m, n be the angles which the moment axis of the resultant, called the *resultant axis*, makes with the co-ordinate axes z, y, x through the centre.

Then the principle of the *parallelopipedon of moments* gives

$$Rp = \sqrt{L^2 + M^2 + N^2}. \quad \ldots \ldots \quad (42)$$

That is, *the resultant moment of a system of forces with respect to any assumed centre of moments is equal to the square root of the sum of the squares of the sums of the moments of the forces of the system with respect to any set of rectangular axes through the assumed centre.* Rp is constant, but L, M and N will, in general, change with the axes.

For the rectangular components of Rp we have

$$\left. \begin{array}{l} Rp \cos n = N; \\ Rp \cos m = M; \\ Rp \cos l = L. \end{array} \right\} \quad \ldots \ldots \quad (43)$$

That is, *the algebraic sum of the moments of the forces of the system with respect to any axis through the assumed centre is equal to the projection of the resultant moment on that axis.*

The position of the resultant axis is given by

$$\left.\begin{aligned} \cos n &= \frac{N}{Rp}; \\ \cos m &= \frac{M}{Rp}; \\ \cos l &= \frac{L}{Rp}. \end{aligned}\right\} \quad \ldots \ldots \quad (44)$$

38. Parallel Forces.—Assuming a system of parallel forces, we have

$$a = \alpha' = \alpha'' = \text{etc.};$$
$$b = \beta' = \beta'' = \text{etc.};$$
$$c = \gamma' = \gamma'' = \text{etc.};$$

and casting out the common factor from Eq. (24), we get

$$R = \Sigma I, \quad \ldots \ldots \ldots \quad (45)$$

the intensities of those forces whose direction cosines are positive being multiplied by $+ 1$, and those whose direction cosines are negative by $- 1$. The intensity of the resultant of a system of parallel forces is therefore equal to the sum of those intensities which act in one direction diminished by the sum of those which act in the opposite direction. The direction of the resultant is that of the intensities whose sum is the greater.

The point of application of the resultant is found from the principle that the moment of the resultant is equal to the sum of the moments of its components. Resuming Eqs. (40) we have, for parallel forces,

$$\left.\begin{aligned} (Rx - \Sigma Ix') \cos b &= (Ry - \Sigma Iy') \cos a; \\ (Rz - \Sigma Iz') \cos a &= (Rx - \Sigma Ix') \cos c; \\ (Ry - \Sigma Iy') \cos c &= (Rz - \Sigma Iz') \cos b. \end{aligned}\right\} \quad \ldots \quad (46)$$

Since these equations must be satisfied for all possible values of a, b and c, the principle of indeterminate coefficients applies, and we have

$$\left.\begin{array}{l} Rx - \Sigma Ix' = 0; \\ Ry - \Sigma Iy' = 0; \\ Rz - \Sigma Iz' = 0. \end{array}\right\} \quad \ldots \ldots \quad (47)$$

Hence

$$\left.\begin{array}{l} x = \dfrac{\Sigma Ix'}{R}; \\ y = \dfrac{\Sigma Iy'}{R}; \\ z = \dfrac{\Sigma Iz'}{R}. \end{array}\right\} \quad \ldots \ldots \quad (48)$$

These values are independent of the angles a, b, c, and will be the same no matter what be the direction of the parallel forces. The point defined by Eqs. (48) is therefore called the *centre of the system*. The position of the centre depends on the intensities of the components and resultant, and upon the points of application of the components.

If the points of application of the component forces be in the same plane, as xy, then $z = 0$, and the centre of the system is in that plane. If the points of application be on the same right line, as the axis of x, then $y = z = 0$, and the centre is on the right line also.

39. Assume a system of two parallel forces. If we take the moments of the components with respect to a point on their resultant, the sum of these moments must be zero since the moment of the resultant with respect to this point is zero. If the two forces act in the same direction, the sum of their moments can be zero only with respect to some point between their action-lines, and we have, Fig. 14,

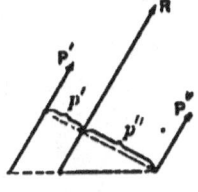

FIG. 14.

$$I'p' - I''p'' = 0; \quad \ldots \ldots \quad (49)$$

$$I' : I'' :: p'' : p'; \quad \ldots \ldots \quad (50)$$

$$\left. \begin{array}{l} I' + I'' : I' :: p'' + p' : p'' :: R : I'; \\ I' + I'' : I'' :: p'' + p' : p' :: R : I''. \end{array} \right\} \ldots (51)$$

40. When the forces act in opposite directions the sum of their moments can be zero only with respect to some point outside the components and nearer the greater, and we have, Fig. 15,

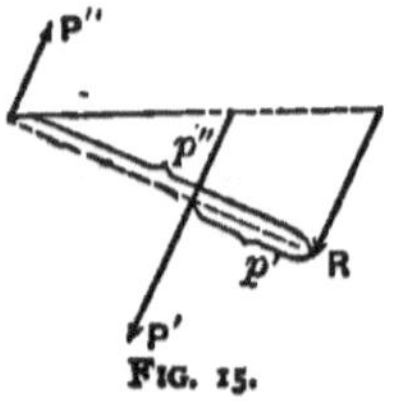

$$I'p' - I''p'' = 0; \quad \ldots \quad (52)$$

$$I' : I'' :: p'' : p'; \ldots \quad (53)$$

Fig. 15.

$$\left. \begin{array}{l} I' - I'' : I' :: p'' - p' : p'' :: R : I'; \\ I' - I'' : I'' :: p'' - p' : p' :: R : I''. \end{array} \right\} \ldots (54)$$

We conclude, therefore,

(1) That the intensities of the components are inversely as the distances of their action-lines from that of the resultant.

(2) That the intensity of either component is to that of the resultant as the distance of the action-line of the other component from that of the resultant is to the distance between the action-lines of the components.

(3) That the intensities of any two of the three forces are inversely as the distances of their action-lines from that of the other.

(4) The resultant lies nearer the component of greater intensity.

(5) If three parallel forces be in equilibrio, the intensities of any two are inversely as the distances of their action-lines from that of the other.

(6) The force of greatest intensity lies between the other two.

41. *A Couple.*—A couple is a pair of equal parallel forces acting in opposite directions but not immediately opposed. The perpendicular distance between the action-lines of the forces is called the *arm* of the couple. From Eq. (54) we have

$$p'' = \frac{l'(p'' - p')}{l' - l''}. \quad \ldots \quad \ldots \quad (55)$$

Let l'' approach l' in value, $p'' - p'$ remaining constant. Then as $l' - l''$ diminishes, p'' increases, and when the two forces become a couple p'' becomes infinite, and R becomes zero. Hence the resultant of a couple is a force of zero intensity at infinity; that is, no single force can replace a couple.

42. The moment of a couple with respect to any centre in the plane of the forces is equal to the product of the arm of the couple by the intensity of one of the forces. To show this let the couple be as indicated in Fig. 16, and assume the three points o, o', o'' as centres. Then we have

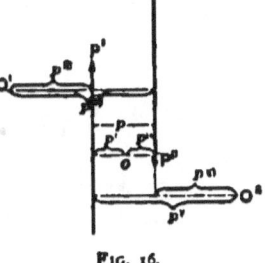

$$\left. \begin{array}{l} l'p' + l''p'' = l'(p' + p'') = l'p; \\ l''p^{\text{iv}} - l'p''' = l'(p^{\text{iv}} - p''') = l'p; \\ l'p^{\text{v}} - l''p^{\text{vi}} = l'(p^{\text{v}} - p^{\text{vi}}) = l'p. \end{array} \right\} (56)$$

Fig. 16.

A couple is represented by its moment, and hence couples may be combined in the same manner as moments.

Gravity.

43. The attraction of the earth for any part of its own mass is called *Gravity;* it is a special case of *universal gravitation*.

The *weight* of a body is the resultant of all the forces of gravity acting on its molecules, and it will be known when its intensity, action-line, and point of application are known. The weight may act either as a stress or as a motive force. In the former case each molecule presses on the one below, so that each horizontal stratum of molecules of a body at rest is subjected to the stress arising from the weight of all above it, and the body is subjected to a compressive strain. When the body is free to move, the weight of each molecule causes its own acceleration, and none of the weight acts as a stress.

44. From geodetic and astronomical observations the earth is found to be an oblate spheroid whose polar semi-axis is approximately 3949.55 ± miles, and equatorial radius 3962.72 ± miles. Were there no rotation of the earth about its axis, the *apparent* weight of a molecule would be that due to the earth's attraction, and it would be directed to the earth's centre. But owing to this rotation the molecule is constantly carried along the circumference of its circle of latitude, and this can only be the case when a force normal to the tangent and directed towards the centre of the circle acts upon it with an intensity $\dfrac{mv^2}{\rho}$ (Arts. 16 and 23).

Let a, Fig. 17, be the molecular mass, al the plane of its circle of latitude, mg its weight, and $ap = \dfrac{mv^2}{\rho}$ the force which is just sufficient to cause it to continue on the circumference of the circle of latitude. We see that ap is that component of the weight of m employed in deflecting it from its rectilinear path, and the other component alone causes pressure on what supports m. This component, mg', is called the *apparent weight*, and is not directed towards the centre of the earth except when m is at one of the poles or on the equator. The maximum value of $\dfrac{mv^2}{\rho}$ due to the angular velocity of the earth is only about $\tfrac{1}{289}$ that of mg; we may therefore, for the present, neglect its consideration and assume that the direction of the weight is towards the earth's centre.

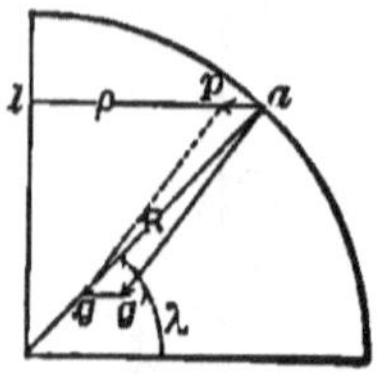

Fig. 17.

45. Again, since the longest dimension of any body whose weight is to be found is insignificant compared with the earth's radius, the action-lines of the molecular weights of any body may be taken parallel to each other, and the system of molecular weights, therefore, to be a system of parallel forces.

From Eq. (2) we may write

$$G = \frac{(M-m)m}{r^2}\mu, \quad \ldots \quad (57)$$

and considering M to be the mass of the earth and m that of the body whose weight is to be found, we see that, owing to the great relative mass of the earth as compared with m, in all practical applications $M - m$ may be taken as a constant, and the intensity of the weight will therefore vary directly with m, and inversely as r^2. Since the attraction is mutual the mass m attracts the earth with the same intensity that the earth attracts m.

The variation in the weights of unit masses in any body at the same locality, due to their increased distances from the centre of the earth, can be neglected. For, assuming the difference of distance to be one mile, the radius of the earth being taken as 4000 miles, the weights of the same body will be as $(4001)^2 : (4000)^2$ or $1 : 1.0005$; that is, there is an increase or diminution of its weight by $\frac{1}{2000}$ part due to a decrease or increase of a mile in distance from the centre of the earth. Therefore the weights of the different unit masses of all bodies whose weights are to be found may be taken to be sensibly equal to each other.

46. The effect of gravity on a free body is to cause it to fall toward the earth with a constantly increasing velocity. Representing the weight of the body by w, and its mass by m, we have (Eq. 20)

$$w = m\frac{d^2s}{dt^2}, \quad \text{or} \quad \frac{d^2s}{dt^2} = \frac{w}{m}. \quad \ldots \quad (58)$$

From this we see that since the ratio of the weights of all bodies to their masses is a constant at the same place on the earth's surface, their accelerations caused by the earth's attraction is also a constant at the same place.

This acceleration is called *the acceleration due to gravity*, and is represented in the text by g. Its value has been determined by much careful experiment, and is given by the equation

$$g = 32.173 - 0.0821 \cos 2\lambda - 0.000003h, \quad \ldots \quad (59)$$

in feet per second, or

$$g = 980.6056 - 2.5028 \cos 2\lambda - 0.000003h, \quad \ldots \quad (60)$$

in centimeters per second, in which λ is the latitude of the place and h the height above the sea-level.

47. The unit of mass depends on the unit of intensity and is, by definition, that quantity of matter which acquires a unit of velocity when acted upon by a force of unit intensity for a unit of time. The weight of the pound mass being the unit of intensity in ordinary use, it is necessary to determine the weight of the corresponding unit of mass. When one pound intensity acts on one pound mass, which is the case when a body falls freely in vacuo, the acceleration is g feet per second; when one pound intensity acts on one unit of mass the acceleration is one foot per second. Since the intensity is the same in both cases, its measure, the product of the mass by the acceleration, is constant; hence we have

$$1 \text{ lb.} \times g = g \text{ lbs.} \times 1. \quad \ldots \quad (61)$$

But the quantity of matter in the second case is the unit of mass, and the unit of mass therefore weighs g lbs. Hence we have for the weight of any body

$$w = mg. \quad \ldots \quad (62)$$

48. *The Centre of Gravity.* —The Centre of Gravity of a body is that point through which the action-line of the body's weight always passes.

Let m and w be the type-symbols for the masses and weights of the molecules of a body, g the weight of the unit mass, and x, y, z the co-ordinates of m. Then, since the action-lines of the molecular weights may be taken parallel, the system becomes a system of parallel forces whose centre, which is *the centre of gravity*, is given by Eqs. (48):

$$\left. \begin{array}{l} \bar{x} = \dfrac{\Sigma wx}{\Sigma w} = \dfrac{\Sigma mgx}{\Sigma mg}; \\[4pt] \bar{y} = \dfrac{\Sigma wy}{\Sigma w} = \dfrac{\Sigma mgy}{\Sigma mg}; \\[4pt] \bar{z} = \dfrac{\Sigma wz}{\Sigma w} = \dfrac{\Sigma mgz}{\Sigma mg}. \end{array} \right\} \quad \ldots \ldots \quad (63)$$

Assuming $g = g' = g'' =$ etc., Eqs. (63) become

$$\bar{x} = \frac{\Sigma mx}{M}; \quad \bar{y} = \frac{\Sigma my}{M}; \quad \bar{z} = \frac{\Sigma mz}{M}. \quad \ldots \quad (64)$$

The point defined by Eqs. (64) is called the *centre of mass*. Therefore, g being considered constant, the centre of gravity is at the centre of mass. The centre of gravity is, accurately, a little below the centre of mass, but in ordinary bodies the distance between them is negligible.

From Eqs. (64) we see that *the product of the mass of the body by either co-ordinate of the centre of mass, referred to any origin whatever, is equal to the algebraic sum of the products of all the molecular masses by their corresponding co-ordinates referred to the same origin*. This is called the *principle of the centre of mass*.

Substituting for m, m', m'', etc., their values in terms of volume and density, Eq. (1), we have

$$\bar{x} = \frac{\Sigma v\delta x}{\Sigma v\delta}; \quad \bar{y} = \frac{\Sigma v\delta y}{\Sigma v\delta}; \quad \bar{z} = \frac{\Sigma v\delta z}{\Sigma v\delta}. \quad \ldots \quad (65)$$

If the body be homogeneous, then $\delta = \delta' = \delta'' =$ etc., and Eqs.

(65) become

$$\bar{x} = \frac{\Sigma vx}{V}; \quad \bar{y} = \frac{\Sigma vy}{V}; \quad \bar{z} = \frac{\Sigma vz}{V}. \quad \ldots \quad (66)$$

That is, in homogeneous bodies the centre of gravity coincides with the centre of volume.

When a body is of such a form that we may obtain expressions for the relations between its surface co-ordinates, and its density is a function of these co-ordinates, we may write

$$\left.\begin{aligned}\bar{x} &= \frac{\Sigma mx}{\Sigma m} = \frac{\int x dM}{M} = \frac{\int \delta x dV}{\int \delta dV}; \\ \bar{y} &= \frac{\Sigma my}{\Sigma m} = \frac{\int y dM}{M} = \frac{\int \delta y dV}{\int \delta dV}; \\ \bar{z} &= \frac{\Sigma mz}{\Sigma m} = \frac{\int z dM}{M} = \frac{\int \delta z dV}{\int \delta dV};\end{aligned}\right\} \quad \ldots \quad (67)$$

and if δ be constant we shall have

$$\left.\begin{aligned}\bar{x} &= \frac{\int x dV}{V}; \\ \bar{y} &= \frac{\int y dV}{V}; \\ \bar{z} &= \frac{\int z dV}{V}.\end{aligned}\right\} \quad \ldots \ldots \quad (68)$$

From these equations the co-ordinates of the centre of gravity can be found by integrating between the limits which determine the volume, when the expressions which enter them are integrable.

49. *Determination of the Position of the Centre of Gravity.*— The magnitudes whose centres of gravity are to be found are supposed to be homogeneous bodies, lines having a uniform cross-section and surfaces uniform thickness. If the body be symmetrical with respect to a plane, this plane may be taken as *xy*, and we have $\bar{z} = 0$; that is, the centre of gravity is in the plane of symmetry. If the body be symmetrical with respect to a right line the line may be taken as the axis of *x*, and we have $\bar{y} = 0, \bar{z} = 0$; that is, the centre of gravity is on the line of symmetry.

50. *Centre of Gravity of Lines.*—The centre of gravity of a *right line* is, by the principle of symmetry, at its middle point.

The centre of gravity of *broken lines* can be found by Eqs. (63) when *x, y, z* are the type-symbols of the co-ordinates of the centre of gravity of each straight portion, and its weight *w* is taken proportional to its length. In this way the centre of gravity of the perimeter of any polygon, or of any number of connected or disconnected right lines, can readily be found.

The differential of any line is

$$dl = \sqrt{dx^2 + dy^2 + dz^2}; \quad \therefore l = \int \sqrt{dx^2 + dy^2 + dz^2}, \quad (69)$$

and therefore Eqs. (68) become, for lines,

$$\left. \begin{array}{l} \bar{x} = \dfrac{\int_{x''}^{x'} x \sqrt{dx^2 + dy^2 + dz^2}}{l}; \\[1ex] \bar{y} = \dfrac{\int_{x''}^{x'} y \sqrt{dx^2 + dy^2 + dz^2}}{l}; \\[1ex] \bar{z} = \dfrac{\int_{x''}^{x'} z \sqrt{dx^2 + dy^2 + dz^2}}{l}. \end{array} \right\} \quad \ldots \quad (70)$$

If the line be a plane curve we may assume xy as its plane, and the above reduce to

$$\left.\begin{aligned}\bar{x} &= \frac{\int_{x''}^{x'} x\sqrt{dx^2+dy^2}}{l}; \\ \bar{y} &= \frac{\int_{x''}^{x'} y\sqrt{dx^2+dy^2}}{l}; \\ \bar{z} &= 0.\end{aligned}\right\} \quad \ldots \ldots (71)$$

Ex. 1. *A Circular Arc.*—Take the axis of y, Fig. 18, as the axis of symmetry, and the origin of co-ordinates at the centre of the arc. Then we have

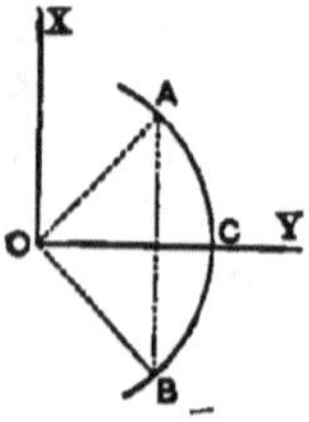

$$\bar{y} = \frac{\int_{-x'}^{+x'} y\sqrt{dx^2+dy^2}}{l};$$

$$x^2 + y^2 = r^2;$$

$$\frac{dy}{dx} = -\frac{x}{y};$$

$$\sqrt{dx^2+dy^2} = dx\sqrt{1+\frac{dy^2}{dx^2}} = dx\sqrt{1+\frac{x^2}{y^2}} = \frac{r\,dx}{y};$$

$$\bar{y} = \frac{\int_{-x'}^{+x'} r\,dx}{l} = \frac{2x'r}{l}.$$

Fig. 18.

Hence the centre of gravity of a circular arc is on its radius of symmetry, and at a distance from the centre of the circle equal to a fourth proportional to the arc, radius and chord.

Ex. 2. *A Cycloid.*—Let x, Fig. 19, be the symmetrical axis; then

$$\bar{y} = 0 \quad \text{and} \quad \bar{x} = \frac{\int x \, dl}{l}.$$

Taking the equation of the cycloid, $l^2 = 8rx$, we have

$$l = 2(2rx)^{\frac{1}{2}} \quad \text{and} \quad dl = (2r)^{\frac{1}{2}} x^{-\frac{1}{2}} dx,$$

whence

$$\bar{x} = \frac{\int_0^x (2r)^{\frac{1}{2}} x^{\frac{1}{2}} dx}{2(2rx)^{\frac{1}{2}}} = \frac{x}{3}$$

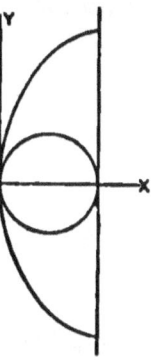

Fig. 19.

Therefore, for the curve corresponding to one complete rotation of the generating circle, $x = 2r$ and $\bar{x} = \frac{2r}{3}$; that is, the centre of gravity is on the axis of symmetry and at a distance from the vertex equal to one third the diameter of the generating circle.

51. *Centre of Gravity of Surfaces.*—For surfaces we have

$$\left. \begin{array}{l} \bar{x} = \dfrac{\int x \, ds}{s}; \\[6pt] \bar{y} = \dfrac{\int y \, ds}{s}; \\[6pt] \bar{z} = \dfrac{\int z \, ds}{s}; \end{array} \right\} \quad \ldots \ldots \quad (72)$$

in which the elementary area ds is given by

$$ds = \frac{dx \, dy}{\cos \beta}; \quad \ldots \ldots \quad (73)$$

and β, the angle which the tangent plane to the surface makes with the plane xy, is given by

$$\cos \beta = \pm \frac{\frac{dL}{dz}}{\sqrt{\frac{dL^2}{dx^2} + \frac{dL^2}{dy^2} + \frac{dL^2}{dz^2}}}; \quad \ldots \quad (74)$$

$$L = f(x, y, z) = 0 \quad \ldots \ldots \quad (75)$$

being the equation of the surface.

52. If the surface be *plane*, we may take it in the plane xy and then we have

$$ds = dx\, dy, \quad \ldots \ldots \ldots \quad (76)$$

and Eqs. (72) become

$$\left.\begin{array}{l} \bar{x} = \dfrac{\int\int x\, dx\, dy}{S}; \\[1em] \bar{y} = \dfrac{\int\int y\, dy\, dx}{S}; \\[1em] \bar{z} = 0. \end{array}\right\} \quad \ldots \ldots \quad (77)$$

Integrating with respect to y, between the limits y' and y'', these become

$$\left.\begin{array}{l} \bar{x} = \dfrac{\int_{x'}^{x'} (y'' - y')x\, dx}{S}; \\[1em] \bar{y} = \dfrac{\frac{1}{2}\int_{x'}^{x'} (y''^2 - y'^2)\, dx}{S}. \end{array}\right\} \quad \ldots \ldots \quad (78)$$

Ex. 1. A Triangle.—If we consider the area of the triangle, Fig. 20, to be made up of right lines drawn parallel to the base *ab*, the weights of these lines act at their middle points, and the centre of this system of parallel forces is somewhere on the line *cd*, drawn from *c* to the middle point of *ab*. Similarly the centre of gravity of the triangle will be found on the lines *ae* and *bf* drawn from the other two vertices to the middle points of their opposite sides. Hence o is the centre of gravity of the triangle. But we know from geometry that the point of intersection o is at two thirds the distance from either vertex to the middle of the opposite side. To show this analytically, let

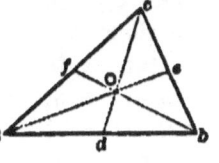

Fig. 20.

$$y'' = \alpha x, \quad y' = \beta x,$$

be the equations of *ac* and *ab*, Fig. 21, the axis of *y* being taken parallel to the side *bc*; then, Eqs. (78),

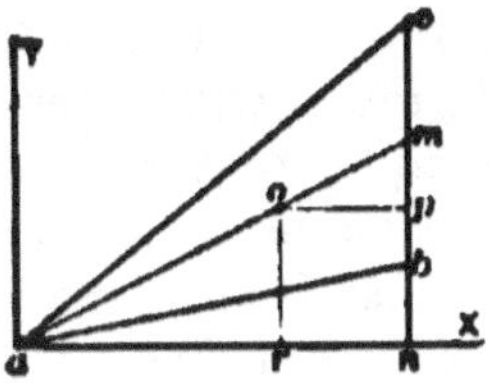

$$\bar{x} = \frac{\int_0^{x'} (\alpha - \beta) x^2 \, dx}{\int_0^{x'} (\alpha - \beta) x \, dx} = \frac{2}{3} x';$$

$$\bar{y} = \frac{\tfrac{1}{2}\int_0^{x'} (\alpha^2 - \beta^2) x^2 \, dx}{\int_0^{x'} (\alpha - \beta) x \, dx} = \frac{2}{3} \frac{(\alpha + \beta)}{2} x';$$

Fig. 21.

but

$$\frac{(\alpha + \beta)}{2} x' = mn \quad \text{and} \quad \frac{2}{3} \frac{(\alpha + \beta)}{2} x' = np = \frac{2}{3} mn.$$

It is evident that a straight rod, in which the weight of each cross-section varies directly as its distance from one extremity, is a precisely similar problem to that of the triangle; and hence its centre of gravity is at two thirds its length from this extremity. Similarly the areas of the bounding surfaces of the sections of a cone or pyramid vary proportionally with their distances from the vertex; hence the centre of gravity of the surface of a cone or pyramid is at two thirds the distance of the base from the vertex.

Ex. 2. *A Polygonal Plane Area.*—Divide the area into triangles, and find their centres of gravity separately; then, by means of the equations

$$\left. \begin{array}{l} \bar{x} = \dfrac{\Sigma w x'}{\Sigma w}, \\ \bar{y} = \dfrac{\Sigma w y'}{\Sigma w}, \end{array} \right\} \quad \cdots \cdots \quad (79)$$

in which w is the type-symbol of the weight of each triangular area, and x', y' the co-ordinates of its centre of gravity, the centre of gravity of the whole area may be found.

Ex. 3. *A Circular Sector.*—A circular sector may be considered as made up of an indefinitely great number of equal triangles having equal altitudes and bases, their vertices being at the centre of the circle. Their centres of gravity will be found on the arc of a circle drawn with a radius equal to two thirds that of the given circle. If the mass of each triangle be supposed concentrated in its centre of gravity, the locus of these masses will be a homogeneous line, and its centre of gravity will coincide with that of the sector. Therefore, calling r the radius of the circular sector, c its chord, and a the length of its arc, we have, taking the axis x to be the axis of symmetry,

$$\bar{x} = \frac{2}{3} \frac{rc}{a}.$$

That is, the centre of gravity of a circular sector is on its radius of symmetry, and at a distance from the centre equal to two thirds of a fourth proportional to the arc, radius and chord.

Ex. 4. *A Circular Segment.*—Let the origin be at the centre of the circle, Fig. 22, and the axis of x that of symmetry; then

$$\bar{x} = \frac{\int_r^{r''}(y''-y')x\,dx}{s}; \quad \ldots \quad (80)$$

$$\bar{y} = 0; \qquad x^2 + y^2 = r^2.$$

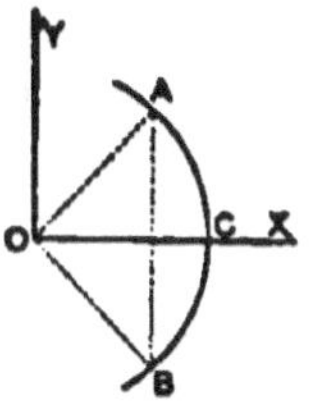

Then we have

$$y'' = (r^2 - x^2)^{\frac{1}{2}}; \qquad y' = -(r^2 - x^2)^{\frac{1}{2}};$$

and Eq. (80) becomes

$$\bar{x} = \frac{2\int_r^{r''}(r^2-x^2)^{\frac{1}{2}}x\,dx}{s} = \frac{\frac{2}{3}(r^2-x''^2)^{\frac{3}{2}}}{s}.$$

But $AB = 2(r^2 - x''^2)^{\frac{1}{2}} = c$; $\therefore \bar{x} = \frac{c^3}{12s}$.

Fig. 22.

Therefore the centre of gravity of a *circular segment* is on the radius drawn to the middle of the arc, and at a distance from the centre equal to the cube of the chord divided by twelve times the area of the segment.

Ex. 5. *An Area bounded by a Parabola, its Axis and one of its Ordinates.*—Take the parabola as in Fig. 23, and we have

$$y''^2 = 2px; \qquad y' = 0;$$

$$\bar{x} = \frac{\int_0^{x''}(y''-y')x\,dx}{\int_0^{x''}(y''-y')dx} = \frac{\int_0^{x''}x^{\frac{3}{2}}\,dx}{\int_0^{x''}x^{\frac{1}{2}}\,dx} = \frac{3}{5}x'';$$

Fig. 23.

$$\bar{y} = \frac{\frac{1}{2}\int_0^{x''}(y''^2 - y'^2)dx}{\int_0^{x''}(y'' - y')dx} = \frac{\frac{1}{2}\int_0^{x''} 2px\, dx}{\int_0^{x''}\sqrt{2px}\, dx} = \frac{3}{8}\sqrt{2px''} = \frac{3}{8}y''.$$

For the parabolic spandrel OBC we have

$$x'' = \frac{y^2}{2p}; \qquad x' = 0;$$

$$\bar{x} = \frac{\frac{1}{2}\int_0^{y''}(x''^2 - x'^2)dy}{\int_0^{y''}(x'' - x')dy} = \frac{\frac{1}{2}\int_0^{y''}\frac{y^4\, dy}{4p^2}}{\int_0^{y''}\frac{y^2}{2p}dy} = \frac{3}{10}\frac{y^2}{2p} = \frac{3}{10}x'';$$

$$\bar{y} = \frac{\int_0^{y''}(x'' - x')y\, dy}{\int_0^{y''}(x'' - x')dy} = \frac{\int_0^{y''}\frac{y^3\, dy}{2p}}{\int_0^{y''}\frac{y^2}{2p}dy} = \frac{3}{4}y''.$$

These latter values can be readily determined by the application of the principle of moments; thus, the area

$$OBA = \tfrac{2}{3}OCBA, \quad \text{and} \quad OBC = \tfrac{1}{3}OCBA.$$

The sum of the moments of the weights of OBA and OBC with respect to O must be equal to the moment of the weight of the rectangle with respect to the same point. Hence we have

$$\frac{2W}{3} \times \frac{3x''}{5} + \frac{W}{3}\bar{x} = \frac{Wx''}{2},$$

and therefore

$$\bar{x} = \frac{3x''}{10};$$

for the x co-ordinate of the centre of gravity of the parabolic spandrel. The y co-ordinate is obtained by considering the figure to be resting on OC, and taking the moments again with respect to O. We have

$$\frac{2W}{3} \times \frac{3y''}{8} + \frac{W}{3}\bar{y} = \frac{Wy''}{2};$$

whence

$$\bar{y} = \frac{3}{4}y''.$$

Ex. 6. *An Elliptic Quadrant.*—The equation of the ellipse referred to its centre and axes is $a^2y^2 + b^2x^2 = a^2b^2$; whence we have

$$y'' = \frac{b}{a}(a^2 - x^2)^{\frac{1}{2}}; \quad y' = 0;$$

$$\bar{x} = \frac{\int_0^a \frac{b}{a}(a^2 - x^2)^{\frac{1}{2}} x\, dx}{\frac{\pi ab}{4}} = \frac{4}{\pi ab}\left(-\frac{b}{3a}(a^2 - x^2)^{\frac{3}{2}}\right)_0^a = \frac{4a}{3\pi};$$

$$\bar{y} = \frac{\frac{1}{2}\int_0^a \frac{b^2}{a^2}(a^2 - x^2)\,dx}{\frac{\pi ab}{4}} = \frac{4b}{3\pi}.$$

If $a = b$ the ellipse becomes a circle, and the co-ordinates of the centre of gravity of a circular quadrant referred to its centre are

$$\bar{x} = \bar{y} = \frac{4r}{3\pi}.$$

53. *Surfaces of Revolution.*—Let x be the axis of revolution; and since it is an axis of symmetry, $\bar{y}$ and $\bar{z}$ are both zero. Let $y = f(x)$ be the equation of the curve whose revolution generates the surface, and we have for the elementary surface

$$2\pi y(dx^2 + dy^2)^{\frac{1}{2}};$$

hence

$$\bar{x} = \frac{\int_{x''}^{x'} 2\pi yx(dx^2 + dy^2)^{\frac{1}{2}}}{\int_{x''}^{x'} 2\pi y(dx^2 + dy^2)^{\frac{1}{2}}} = \frac{\int_{x''}^{x'} yx(dx^2 + dy^2)^{\frac{1}{2}}}{\int_{x''}^{x'} y(dx^2 + dy^2)^{\frac{1}{2}}}.$$

Ex. 1. *A Spherical Zone.*—The equation of the circle is $x^2 + y^2 = r^2$; then we have

$$\bar{x} = \frac{\int_{x''}^{x'} yx(dx^2 + dy^2)^{\frac{1}{2}}}{\int_{x''}^{x'} y(dx^2 + dy^2)^{\frac{1}{2}}} = \frac{\int_{x''}^{x'} rx\, dx}{\int_{x''}^{x'} r\, dx} = \frac{x' + x''}{2}.$$

Whence the centre of gravity of a spherical zone is at the middle point of the right line joining the centres of its bases.

Ex. 2. *A Right Conical Surface.*—Let the equation of its generating line be $y = ax$, and we have

$$\bar{x} = \frac{\int_0^{x'} x^2\, dx}{\int_0^{x'} x\, dx} = \frac{2}{3}x'.$$

Hence the centre of gravity of a right conical surface is on its axis and two thirds of the distance from the vertex to the base; it is independent of the angle of the cone, and hence is a common point for all right cones having the same vertex, axis and altitude.

54. *Centre of Gravity of Volumes.*—For volumes we have, in general, $dV = dx\, dy\, dz$, and Eqs. (68) become

$$\left.\begin{aligned} \bar{x} &= \frac{\iiint x\, dx\, dy\, dz}{V}; \\ \bar{y} &= \frac{\iiint y\, dx\, dy\, dz}{V}; \\ \bar{z} &= \frac{\iiint z\, dx\, dy\, dz}{V}. \end{aligned}\right\} \quad \ldots \ldots (81)$$

55. *Volumes of Revolution.* — A volume of revolution being symmetrical with respect to its axis, if we take this axis as the axis of x we shall have, Eqs. (68),

$$\left.\begin{aligned} \bar{x} &= \frac{\int x\, dV}{V} = \frac{\int_{x''}^{x'} \pi y^2 x\, dx}{\int_{x''}^{x'} \pi y^2\, dx} = \frac{\int_{x''}^{x'} y^2 x\, dx}{\int_{x''}^{x'} y^2\, dx}; \\ \bar{y} &= 0; \\ \bar{z} &= 0. \end{aligned}\right\} \quad \ldots (82)$$

Ex. 1. *A Paraboloid of Revolution.* — To find the centre of gravity of a portion of a paraboloid of revolution limited by planes perpendicular to the axis, we have for the equation of the generating curve

$$y^2 = 2px,$$

and for the co-ordinate of the centre of gravity

$$\bar{x} = \frac{\int_{x''}^{x'} 2px^2\, dx}{\int_{x''}^{x'} 2px\, dx} = \frac{2(x'^3 - x''^3)}{3(x'^2 - x''^2)}.$$

When $x'' = 0$ we have

$$\bar{x} = \tfrac{3}{4}x.$$

Ex. 2. *A Spheroid.* — Taking the origin at the extremity of the axis, the equation of the generating curve is

$$y^2 = \frac{b^2}{a^2}(2ax - x^2),$$

and for a portion of the spheroid from the origin to x' we have

$$\bar{x} = \frac{\int_0^{x'}(2ax - x^2)x\,dx}{\int_0^{x'}(2ax - x^2)dx} = \frac{x'}{4}\frac{(8a - 3x')}{(3a - x')}.$$

For half of the spheroid we have $x' = a$, and

$$\bar{x} = \tfrac{5}{8}a;$$

which is independent of the shape of the spheroid.

56. Whenever the volumes whose centres of gravity are to be found are such that we can connect the areas of their successive sections normal to any line by some law, their centres of gravity can be found from the general equations by a single integration. Thus to find the centre of gravity of any cone or pyramid, first find the centre of gravity of its base and join this point with the vertex. It is evident that the line so drawn will pierce the successive sections in their centres of gravity, and at these points the weights of the several sections will act, with intensities which are proportional to their areas. Then Eqs. (68), when X is the area of any section parallel to the base, become

$$\bar{x} = \frac{\int_{x''}^{x'} Xx\,dx}{V}; \quad \ldots \ldots \quad (83)$$

the line from the vertex perpendicular to the base being taken as the axis of x.

Ex. *A Pyramid or Cone.*—Take the origin at the vertex, and let A be the area of the base and x' the abscissa of its centre of gravity. Then we have for any section

$$X = A\frac{x^2}{x'^2},$$

and

$$\bar{x} = \frac{\frac{A}{x'^2}\int_0^{x'} x^3\, dx}{\frac{A}{x'^2}\int_0^{x'} x^2\, dx} = \frac{3}{4}x'.$$

Therefore the centre of gravity of a pyramid or cone is on the line joining the vertex with the centre of gravity of the base, and at a distance from the vertex equal to three fourths of its length.

57. *Theorems of Pappus.*—Clearing the second of Eqs. (71) and (78) of fractions and multiplying both members of each by 2π, we have

$$2\pi \bar{y} l = \int_{x''}^{x'} 2\pi y\, dl. \quad \cdots \cdots \cdots \quad (84)$$

$$2\pi \bar{y} s = \int_{x''}^{x'} \pi(y'''^2 - y''^2)\, dx. \quad \cdots \cdots \quad (85)$$

The second member of Eq. (84) is the expression for the area of the surface of revolution generated by the curve l about the axis x, and that of Eq. (85) is the expression for the volume generated by the revolution of the plane surface s about the same axis. Hence we have by these equations a simple means of determining an area or volume of revolution whenever we can find the position of the centre of gravity of the generating line or surface. For by the first members we see that such area or volume is equal to the product obtained by multiplying the length of the generating line, or the area of the generating plane surface, by the circumference described by its centre of gravity. These theorems are useful in mensuration.

Graphical Statics.

58. Roof and bridge trusses are usually framed structures composed of beams united by rods and struts, and are subjected to certain stresses due to the weights of the assembled parts, the loads they are required to support, and wind pressure. While the computation of these stresses can be made by the usual analytical methods, the processes of *graphical statics* are so simple and accurate as to make them of frequent application. The following pages contain merely an exposition of its simplest fundamental principles, to which are added a few of the more elementary illustrative examples.* The further development more properly belongs to applied mechanics in Civil Engineering.

59. *Reciprocal Figures.*—Any two figures are said to be reciprocal when the first can be derived from the second in the same way that the second is obtained from the first. Reciprocal figures applicable to graphical statics are subject to the following conditions :

(1) The sides of one should be respectively perpendicular or parallel to those of the other.

(2) Lines radiating from a vertex in one figure should be perpendicular or parallel to corresponding sides forming a closed polygon in the other.

(3) Each figure should be composed of the same number of closed polygons, and each line of the figure should make a part of two of these polygons, and of two only.

---

* La Statique Graphique et ses applications aux construction, par Maurice Levy. Gauthier-Villars; Paris, 1874.

(4) At least three lines should meet at each vertex, and each side should pass through at least two vertices.

The two figures $DABC$ and $dabc$ (Fig. 24), each formed by the six lines joining four points in a plane, fulfil all the above conditions, and are therefore reciprocal figures applicable to graphical statics. Either of these figures is completely determined when any five of its six lines are given; for, any five of the lines form two triangles having a common side, and the sixth line joins two vertices which are already fixed in position by the five given lines. Hence if all the lines in two figures, except one in each, are known to fulfil the conditions of reciprocal figures, the figures are reciprocal.

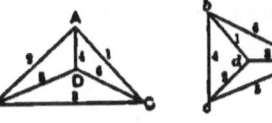

FIG. 24.

60. *The Force Polygon.*—Let there be a given system of coplanar forces; then if, from any point in the plane, a polygon be constructed whose sides taken in order represent the directions and intensities of the several forces of the system, this polygon is called a *force polygon* of the system. By the principle of the parallelogram of forces we readily see that, if the polygon be closed, the resultant of the system is zero and the forces are in equilibrio; and in general, that the right line required to close the polygon, *when reversed*, represents the resultant of the system in *intensity* and *direction*.

61. *The Polar Polygon.*—Let the polygon *abcde*, Fig. 25, be the force polygon of the system of forces 1 1′, 2 2′, 3 3′, 4 4′, Fig. 26. Assume any point O in its plane as a pole, and draw the lines Oa, Ob, Oc, etc., to its vertices. In the plane of the forces draw any line RR'' parallel to Oa of the force polygon, that is, parallel to the line joining the pole and the origin of the side 1 of the force polygon, and mark its intersection with the force 1 1′ by the symbol 1. From 1 draw 1 2 parallel to Ob, and mark its intersection with the force 2 2′ by 2. In the same way fix the points 3, 4 and R; the latter point being the intersection of the line drawn through 4 parallel to Oe with the first line drawn. The polygon 1 2 3 4 R1 is called a *funicular* or *polar polygon* of the system of forces, with reference to the pole O. Since $1R$ might have occupied any position parallel to Oa, it is determinate in *direction* but arbitrary in *position;* therefore the polygon constructed is but one of an indefinitely great number of polar

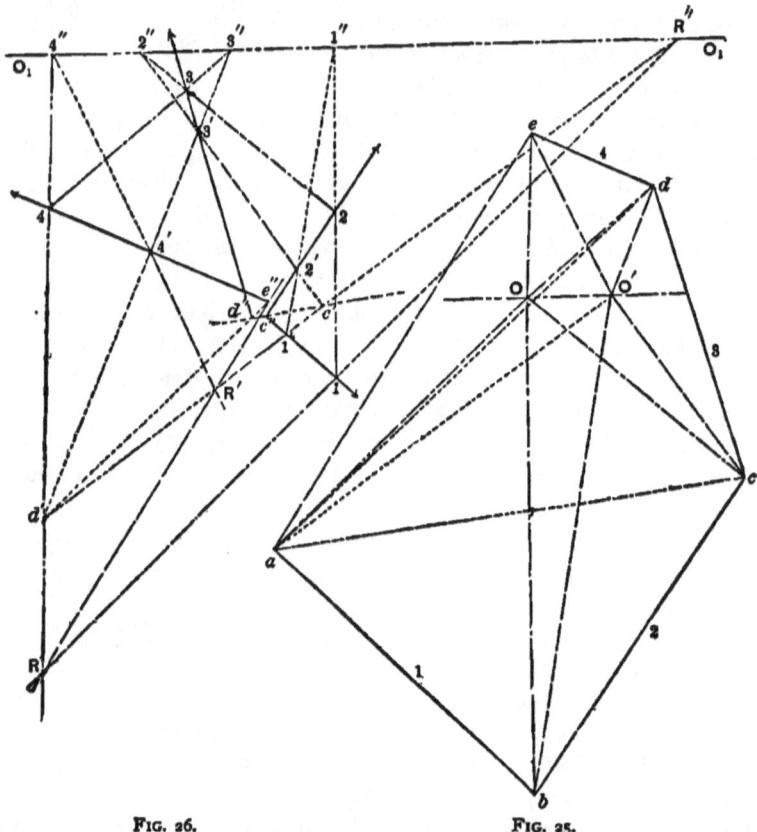

FIG. 26.   FIG. 25.

polygons belonging to the pole O. All of these, however, having parallel sides, are similar figures. Assuming another pole, O', and line $R'R''$ parallel to $O'a$, the polar polygon $1'\ 2'\ 3'\ 4'\ R'$ may be constructed, belonging to another set of similar polygons; and so on indefinitely. From their construction it is evident that the force polygon and the polar polygon are reciprocal figures.

62. *Properties of Polar Polygons.*—The following properties of polar polygons are of use in graphical solutions:

(1) *The intersections of the corresponding sides of two polar polygons belonging to the same system of forces are on a right line parallel to the line joining the two poles.* To show this, produce the corresponding sides of the two polar polygons, 1 2 and 1' 2', 2 3 and 2' 3', etc., until they intersect; the points 1", 2", 3", 4", R'' will lie on the same right line $O,O,$ drawn through the point of intersection R'' of 1R and 1'R', and parallel to OO'; then five of the six lines joining the points R''', 1, 1' and 1", viz., 1 1', 1 R''', 1 1", 1' R''', 1' 1", of the polar polygon are respectively parallel to the five lines ab, Oa, Ob, $O'a$, $O'b$, joining the four points O, O', a, b, of the force polygon; therefore the sixth lines, R'''1" and OO', of these reciprocal figures are also parallel. In the same way the other points, 2", 3", 4", may be shown to be on the right line through R'' parallel to OO'. Hence the change from the pole O to O' is equivalent to supposing that each of the sides of the first polar polygon rotates around each point of intersection until it coincides with the corresponding side of the second polar polygon.

As the pole O' approaches O, the vertices 1 and 1' of the two polar polygons remaining fixed, the line $O,O,$ moves toward infinity, always remaining parallel to OO'. When O' coincides with O, the sides of the polar polygons become parallel and they become two of the same set, and $O,O,$ is at an infinite distance. Hence the parallelism of polar polygons relative to the same pole is merely a particular case of polar polygons relative to different poles.

(2) *If a system of forces and one of its polar polygons be given, every other polar polygon of the system may be constructed without the aid of the force polygon.* Thus, suppose that the one relative to O be known. Draw the arbitrary line $O,O,$, and prolong the sides of the known polygon to meet this line at 1", 2", 3", etc.; then draw through any point thus determined, as R'', a line ar-

bitrary in direction meeting the force 1 1′ in 1′; from 1′ draw 1′ 1″, and where it meets the force 2 2′ will be the vertex 2′; from 2′ the line 2′ 2″ will determine 3′ by its intersection with the force 3 3′, and so on. To show that this polygon 1′ 2′ 3′ 4′ R′ 1′ is a polar polygon, draw through the origin a of the force polygon the right line aO′ parallel to 1′ R″, and through O a parallel to $O_{,} O_{,,}$ and let O′ be their point of intersection. The two figures formed by five of the lines joining the four points R″, 1, 1′, 1″, and O, O′, a, b are parallel and reciprocal, and therefore the sixth lines 1′ 1″ and O′ b are parallel and the figures are reciprocal. In the same way, taking the figures corresponding to the four points 2, 2′, 1″, 2″, and O, O′, b, c, it can be shown that 2′ 2″ and O′ c are parallel and reciprocal, and so on; hence the polygon 1′ 2′ 3′ 4′ R′ 1′, being reciprocal to the force polygon, is a polar polygon with reference to the pole O′.

(3) *The intersection of any two sides of the polar polygon is a point of the resultant of the forces represented by the lines included between the corresponding vertices of the force polygon.* Prolong the sides 1′ R′ and 2′ 3′ till they meet at c′, and the forces 1 and 2 till they intersect at c″. The figures formed by joining O′, a, b, c, and 1′, 2′, c′, c″, are reciprocal; and hence c′ c″ is the action-line of the resultant of 1 and 2, since it is parallel to ac and passes through c″. Similarly d′ d″ is the action-line of the resultant of ac and 3, or of 1, 2 and 3, and R R′ is the action-line of the resultant of the whole system.

Hence to find the resultant of the whole or any part of any system of co-planar forces by graphical construction, draw any polar polygon on the action-lines of the forces whose resultant is to be found; the intersection of the extreme sides will be a point of the resultant, and a line drawn through this point parallel to the closing line of the force polygon of the forces in question will be the action-line of the resultant. The resultant is then completely determined by laying off on this line, in the proper direction, a distance equal to the length of the closing line of the force polygon.

(4) *When a system of co-planar forces is in equilibrio, and the posi-*

tions of the action-lines of three unknown forces are given, the intensities of these unknown forces may be determined by means of the polar polygon. Let the known forces 1, 2, 3 and 4, and the unknown forces 5, 6 and 7, Fig. 27, be a system in equilibrio. Suppose the known forces to be in equilibrio with two of the unknown forces, as 5 and 7, and construct the force polygon *a b c d e g a*. Assume any pole, as *O*, draw *Oa, Ob, Oc, Od, Oe* and *Og*, and con-

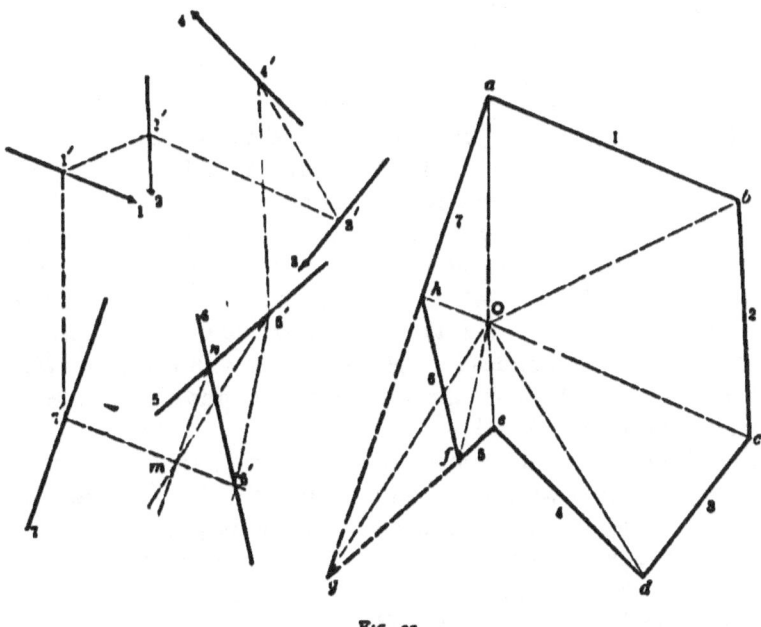

Fig. 27.

struct the polar polygon $7' 1' 2' 3' 4' 5'$. The remaining lines of the polar polygon of the whole system must radiate from $5'$ and $7'$ and intersect on the action-line of the force 6. To determine these lines we must construct a figure which shall be reciprocal to that formed by joining *O, g,* and the two extremities of 6 in the force polygon. Two vertices of the required figure are $5'$ and *n*. The lines *nm, m5'*, and $5'n$ are reciprocal to those radiating from *g*, and *m* is therefore another vertex of the required figure.

Join $7'$ and m, and the intersection of this line with the action-line of the force 6 gives the remaining vertex. The required figure is therefore $mn5'6'$. Now complete the reciprocal of this figure in the force polygon by drawing Of parallel to $5'6'$, and Oh parallel to $7'6'$. Then draw hf, which is parallel to $n6'$ by the reciprocity of the figures, and we have 5, 6 and 7 in the force polygon to represent the intensities and directions of the forces which were to be determined.

63. *Problems.*—Since the triangle is the only polygonal figure that cannot change its form without changing the length of one of its sides, it is made the basis of all frame-work. In the frames here discussed the parts are supposed to be free to rotate about the joints at the vertices of the frame, and they are therefore subjected to longitudinal strains only.

The foregoing principles enable us to find the stresses on the parts of a frame which is a plane figure, when subjected to the action of forces which are co-planar with it. Two diagrams are constructed, one, called *the frame diagram*, to represent the frame and the action-lines of the forces, and the other, *the strain diagram*, to represent the force polygon and the stresses on the various parts.

To facilitate the construction and reading of the strain diagram the following notation is employed: In the frame diagram each triangular space is marked by a letter, and exterior to the frame each space bounded by the action-lines of adjacent forces is also thus marked. Thus, in Fig. 28, A designates the left-hand space of the frame, H the exterior space between forces 1 and 8, and O the exterior space between the forces 7 and 8. Any line or vertex in the figure is designated by the letters of the spaces separated by it. Thus, HO designates the force 8, HI the force 1, and HA the line of the frame between the spaces H and A. The vertex at which force 1 acts may be designated by IA or HB. HO designates both a force and a vertex. In such a case one is called the *force HO*, and the other the *vertex HO*.

In the strain diagram, Fig. 29, the vertices are lettered, and any line of the diagram is designated by the letters at its extremities. The letters are so arranged in the two diagrams that reciprocal lines shall be designated by the same letters. This arrangement is shown in the figures.

(1) The frame represented in Fig. 28 is subjected to the action of the forces 1, 2, 3, 4, 6 and 7, as indicated by their action-lines, the extreme vertices HO and LM being points of support. It is required to find the stresses on the parts of the frame.

To find the reactions at the points of support, construct the force polygon $HIJKLNO$, Fig. 30. HO represents the intensity of the resultant of the applied forces. To find its point of application, assume the pole O', Fig. 30, and construct the polar polygon $R'1'2'3'4'6'7'R'$, Fig. 28. R' is then the point of application of the resultant, and its action-line is therefore $R'R$. Through the points of support draw 5 and 8 parallel to $R'R$, and divide HO, Fig. 30, into two parts which shall be to each other as the distances of R' from the action-lines of 5 and 8. Thus, making OP equal to the distance between the forces 5 and 8, and OQ equal to the distance of R' from 8, and drawing QR parallel to PH, we have OR as the intensity of force 5, and RH that of force 8.

Now construct the force polygon of the whole system $HIJKLMNOH$, Fig. 29, LM being the intensity of 5, and OH that of 8.

To construct the strain diagram begin with a vertex where only two forces are unknown, as the point of support HO. On HO, Fig. 29, construct the reciprocal of this vertex by drawing through H a line parallel to HA, Fig. 28, and through O a line parallel to AO. The triangle $OHAO$ gives the intensities of the three forces acting at the vertex HO. Since OH is the direction of the force 8, the directions of the other two are HA and AO. Also, since HA acts along the frame-piece *towards the vertex* it produces *compression*, and since AO acts *from the vertex* the latter acts as a *tension*.

Passing now to the vertex HB, we have HA and HI known, and drawing AB and IB, Fig. 29, parallel to AB and IB, Fig.

MECHANICS OF SOLIDS.

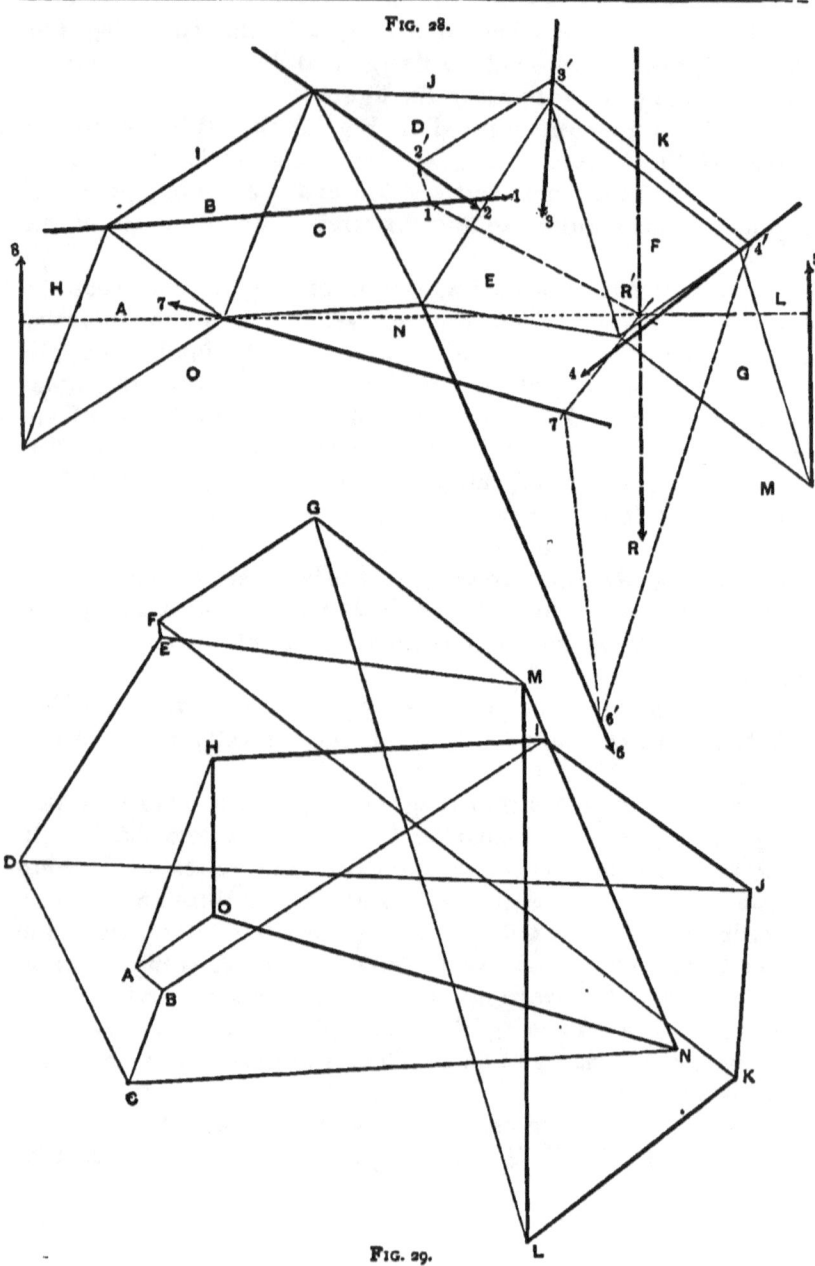

Fig. 28.

Fig. 29.

28, we have the reciprocal of the vertex *HB*, which is *AHIBA*, giving the stresses on the pieces which meet at that vertex. The direction of *HI* is known, and we see by following the

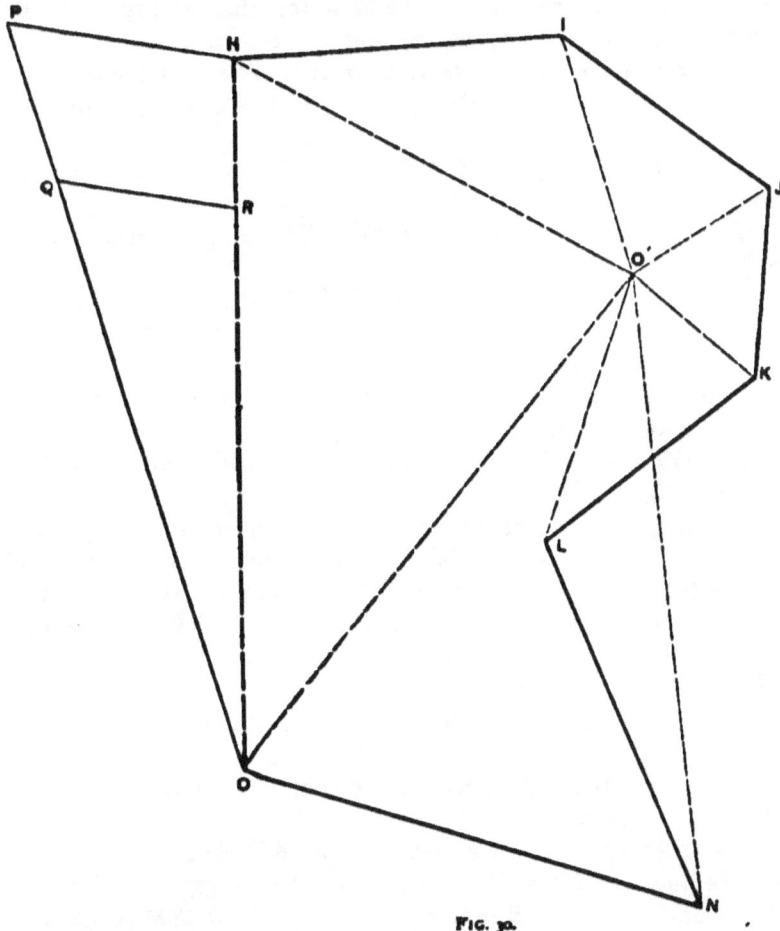

Fig. 30.

strain polygon that *IB*, *BA* and *AH* all act towards the vertex. The stresses on the corresponding pieces are therefore compressions.

For the vertex *AC* we have in the strain diagram *NO*, *OA* and AB. Draw BC and NC parallel to the corresponding lines in the frame, and we have $NOABCN$ for the reciprocal of the vertex AC. BC is compression and CN tension.

Taking the remaining vertices in succession, we have:

For BJ, the polygon $CBIJDC$; JD being compression and DC tension.

For NE, the polygon $MNCDEM$; DE and EM being tensions.

For DK, the polygon $EDJKFE$; KF being compression and FE tension.

For EG, the polygon $MEFGM$; FG and GM being tensions.

For FL, the polygon $GFKLG$; LG being compression.

For LM, the polygon LMG.

From which all the stresses and their characters are completely determined.

(2) *The Simple Warren Truss, having equal loads at the lower vertices* (Fig. 31).

Construct the force polygon of the applied forces $IJKLM$, Fig. 32. Their resultant is IM, which evidently acts through the middle point of the truss. The reactions at the points of support are each equal to half the total load, and are MA and AI.

For the vertex AI we have the polygon $AIBA$; IB being tension and BA compression.

For the vertex IC we have $BIJCB$; JC being tension and BC compression.

For the vertex AC we have $ABCDA$; CD being tension and DA compression.

For the vertex DJ we have $DCJKED$; and for the vertex AE we have $ADEFA$. In the last two polygons, since E and D, and E and F are coincident, the parts DE and EF support no stress.

The stresses on the remaining parts may be found by proceeding with this construction through the vertices KF, AG and LH, or by a construction similar to the above, beginning at the vertex AM.

The upper chord is subjected to a stress of compression and

the lower chord to one of tension, these stresses being greatest at the middle of the truss, while the stresses on the diagonals are greatest at the ends of the truss.

(3) *The Simple Warren Truss, loaded unequally at the lower vertices.*

Let the frame be the same as in the preceding example, Fig.

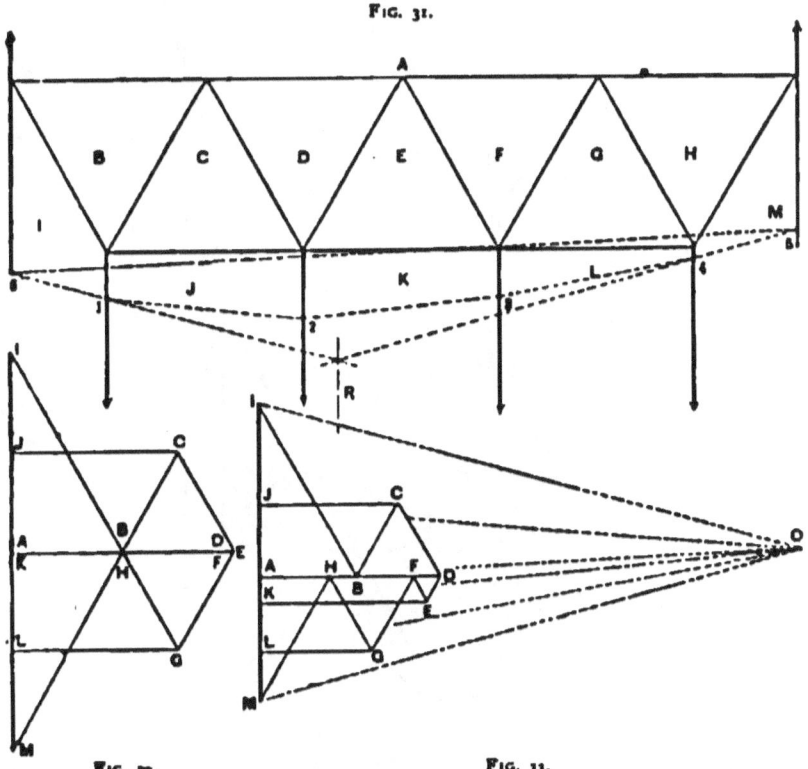

31, and let the two right-hand loads be each one half as great as the others. The force polygon of the applied forces is $IJKLM$ (Fig. 33). The intensity of their resultant is IM. To find its action-line and the reactions at the points of support, assume the pole O and construct the polar polygon 6 1 2 3 4 5 6. R is the action-line of the resultant, and drawing *OA* parallel to 5 6, we have *MA* and *AI* as the intensities of the reactions.

The determination of the strains on the parts is sufficiently indicated by the nomenclature in the diagram.

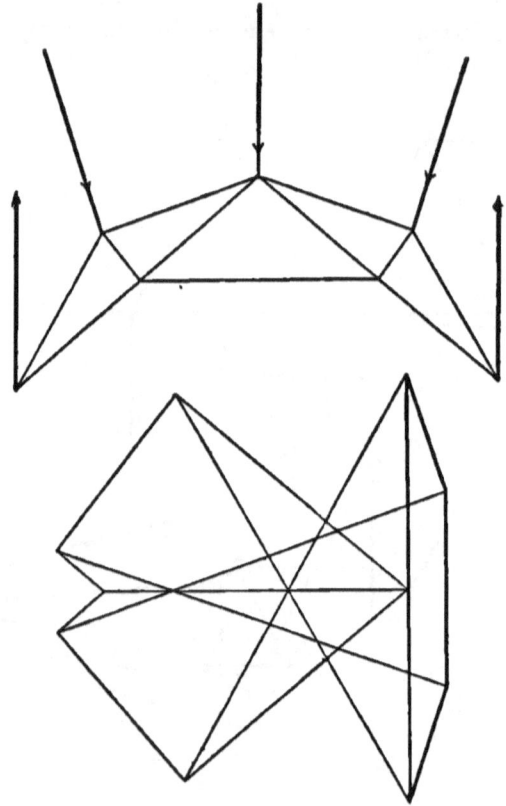

FIG. 34.

(4) Figs. 34, 35, 36 and 37 represent loaded trusses with their strain diagrams. As an exercise the student should supply the nomenclature and determine the character of the strains. It will be observed that at one vertex of the frame in Fig. 37 there are three forces to determine.

GRAPHICAL STATICS. 59

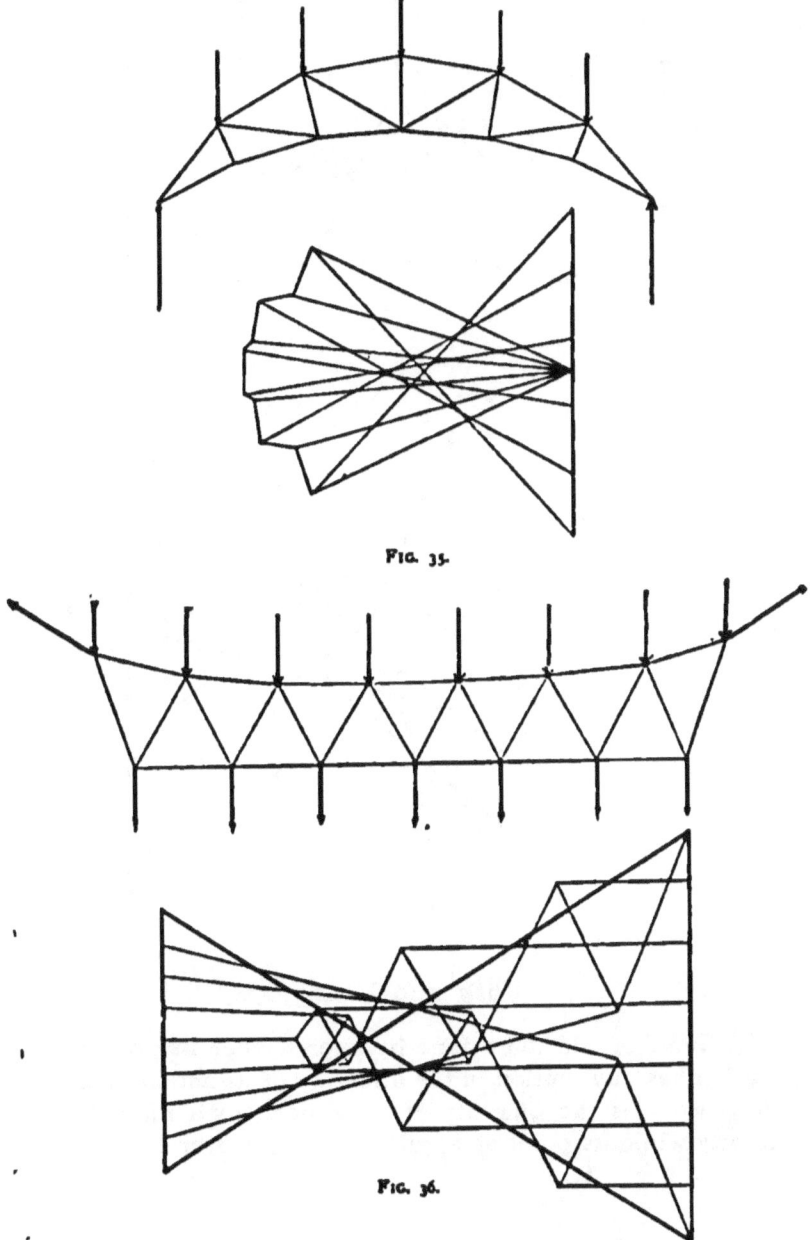

Fig. 35.

Fig. 36.

60   MECHANICS OF SOLIDS.

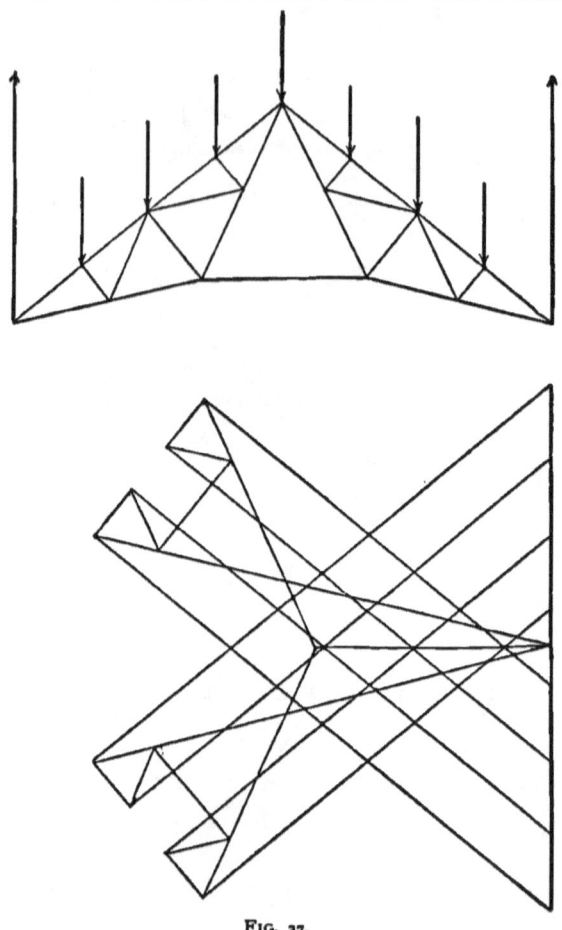

Fig. 37.

Work and Energy.

64. *Work* is said to be done by a force when its point of application has any motion in the direction of its action-line. The unit of work is the quantity of work done by a force of unit intensity while its point of application moves over the distance unity in the direction of the action-line of the force. The unit of work used throughout the text is called the *foot-pound*, the unit of distance being one foot and the unit of intensity one pound. The unit of work of the C. G. S. system is the work done by a dyne over a centimeter, and is called the *Erg*.

The simplest illustration of work is that of lifting a weight through a vertical height. Thus, it requires the expenditure of one foot-pound of work to lift one pound through a height of one foot. Hence, the work done in lifting a weight through any height is equal to the product of the weight and height; and, in general, the work done by any *constant* force is found by multiplying its intensity by the path of its point of application, estimated in the direction of its action-line. If the force be variable it may be regarded as constant while its point of application describes a path dp, *estimated along its action-line*, and the elementary quantity of work is

$$dW = Idp. \qquad (86)$$

The summation of all the elementary quantities of work gives the total quantity done by this force, and we have

$$W = \Sigma dW = \Sigma Idp. \qquad (87)$$

Where the principles of the calculus can be applied, we have

$$W = \int dW = \int Idp. \qquad (88)$$

Hence, whenever such a relation can be established between the variable intensity I and the path p that the second member can be expressed as a known integrable function of a single variable, the total quantity of work can be determined by integration.

65. The symbol expressing work, $\int I dp$, is analogous to the symbol $\int y dx$ in the calculus. The latter is the representative of the quadrature of a curve whose varying ordinates are y and abscissas x, both expressed in the same unit. From this analogy we may graphically represent work by the area contained between the axis of p and a curve whose ordinates measure the varying intensity of the working force at the different points of the path p. The unit of work is graphically represented by a square whose side is the unit of the scale from which the intensity and path are taken.

When the expression $\int I dp$ is not integrable, the quantity of work can be determined approximately by the usual methods for the estimation of the area included between the curve, the extreme ordinates, and the path, as in mensuration. Thus let the ordinates of the curve yy', Fig. 38, represent the varying intensity of the force while its point of application passes over the path pp'. Poncelet's formula for the approximate area is

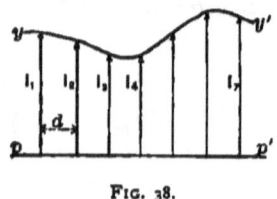

FIG. 38.

$$Q = d[2(I_2 + I_4 + I_6 + \ldots I_n) + \tfrac{1}{4}(I_1 + I_{n+1}) - \tfrac{1}{4}(I_2 + I_n)], (89)$$

in which d is the distance between the consecutive ordinates I_1, I_2, I_3, etc., when the whole path, pp', is divided into any *even* number, n, of equal parts.

If the varying values of I be known only at certain points of the path, the extremities of these may be joined by right lines, thus forming trapezoids whose aggregate area *approximately* represents the quantity of work.

66. *Energy.*—Force and matter are inseparably connected. Any system of masses is accompanied by forces, and these forces perform work during any change in the configuration of the system. Energy is *the capacity for doing work*, and it is measured in units of work. Taking a single molecule of the system moving under the action of the resultant of all the forces applied to it, we have

$$I dp = m \frac{d^2s}{dt^2 \cos \alpha} dp = m \frac{d^2s}{dt^2} ds; \quad \ldots \quad (90)$$

$$\int_2^1 I dp = \int_1^2 m \frac{d^2s}{dt^2} ds = \frac{m(v_2)^2}{2} - \frac{m(v_1)^2}{2}; \quad \ldots \quad (91)$$

and for the whole system,

$$\Sigma \int_2^1 I dp = \frac{\Sigma m(v_2)^2}{2} - \frac{\Sigma m(v_1)^2}{2}. \quad \ldots \quad (92)$$

When the molecule is at the position 1, the quantity of work represented by the first member of Eq. (91) is called *potential energy*, since it measures the capacity of the force to do work while the molecule passes from 1 to 2. It is simply *energy of position*; that is, by virtue of the position of the molecule in the system the forces acting upon it have a certain power to do work while its position is changing. When the molecule arrives at the position 2, the *potential energy* represented by $\int_2^1 I dp$ has been converted into *energy of motion*, called *kinetic energy*, which is measured by $\frac{m(v_2)^2}{2} - \frac{m(v_1)^2}{2}$, the kinetic energy of the molecule being $\frac{m(v_1)^2}{2}$ at 1 and $\frac{m(v_2)^2}{2}$ at 2.

Hence, *potential energy* is defined to be that part of the energy of a system which it possesses by virtue of the relative positions of its different masses, and *kinetic energy* to be the energy which the system possesses by virtue of the motions of its different masses. The term *work* is applied to the *change* of energy from one form or body to another.

For example, let the system be composed of *a unit of mass* and *the earth*, and let the limits be determined by two horizontal planes separated by a distance of 10 feet. Taking the body at the upper limit, we have

$$\int_2^{\prime\prime} I dp = 32.2 \times 10 = 322 \text{ ft.-lbs.}$$

If the body start from rest and fall freely in vacuo to the lower plane, we shall have this *potential energy* converted into *kinetic energy*, or

$$\frac{mv^2}{2} = \frac{v^2}{2} = 322$$

while at any intermediate point, part of the energy is potential and part kinetic. Thus, when the body has fallen to a point midway between the two limits, its potential energy with respect to the lower plane is 161 ft.-lbs., and its kinetic energy is also 161 ft.-lbs. Each form of energy is measured in units of work, but *no work is done unless there be a transformation of energy*. This illustrates what is meant by *energy of position* and *energy of motion*.

To illustrate further, the muscular potential energy in a man's arm may be changed into potential energy of elasticity in a bent bow, and the potential energy of the bow may be changed into kinetic energy of a moving arrow, work being done in both cases. Kinetic energy cannot, however, be transferred from one body to another without passing through the potential form.

67. *The Law of the Conservation of Energy.*—Scientific investigation points to the conclusion that the total quantity of *energy* in the universe, as well as the total quantity of matter, is invariable; that is, that neither matter nor energy can be created or destroyed by any known means. Accepting this as a scientific truth, we must admit that the energy gained or lost in any limited system of masses in which the energy varies must have been obtained from other masses or transferred to them.

A conservative system is one containing a certain definite amount of energy. It consists of limited masses subjected to the action of definite forces. The law of energy for such a system is

$$\Pi + K = C. \quad \ldots \ldots \ldots (93)$$

in which Π represents the potential and K the kinetic energy at any time, and C the constant quantity of energy in the system. Π and K may both vary with the time, but C is constant; and if any change occur in the potential energy we shall have a corresponding and equal but opposite change in the kinetic energy; thus,

$$\Pi_{t'} - \Pi_{t} = K_{t'} - K_{t}, \quad \ldots \ldots (94)$$

each member representing the change in the corresponding energy during the time $t' - t$. During this time the forces of the system act upon the masses and cause them to change their conditions of motion and relative positions, the change in the potential energy being

$$\Sigma \int_{t'}^{t} I d\phi. \quad \ldots \ldots \ldots (95)$$

The change in the kinetic energy of a single molecule of the system during this interval is

$$\int_{t'}^{t} m \frac{d^2 s}{dt^2} ds; \quad \ldots \ldots \ldots (96)$$

and for the whole system,

$$\Sigma \int_{t'}^{t} m \frac{d^2 s}{dt^2} ds. \quad \ldots \ldots \ldots (97)$$

The change in the potential energy in the time dt is evidently

$$\Sigma I d\phi, \quad \ldots \ldots \ldots (98)$$

and the change in the kinetic energy during the same time is

$$\Sigma m \frac{d^2 s}{dt^2} ds. \quad \ldots \ldots \ldots \quad (99)$$

Since these two quantities are always equal, we have

$$\Sigma I dp = \Sigma m \frac{d^2 s}{dt^2} ds; \quad \ldots \ldots \quad (E)$$

an equation which expresses the *Law of the Conservation of Energy*. This law may be stated as follows:

The total energy of any conservative system is a quantity which cannot be increased or diminished by any mutual action of the bodies of the system, and any change of either potential or kinetic energy must always be accompanied by an equal change in the other.

It is evident from this statement that the universe is the only rigidly conservative system. But many limited systems are so remote from all other bodies that the effect of these latter is insignificant when considering the relative motions of the former.

Eq. (E) is the *fundamental equation of mechanics*, and it involves all relative changes in the configuration and motion of any conservative system.

68. *The Principle of Virtual Velocities.*—If no change of state occur in any of the molecules, the factors $\frac{d^2 s}{dt^2}$ will each become zero, and the equation reduces to

$$\Sigma I dp = 0; \quad (S)$$

or, the total quantity of work done by the forces upon the system of masses is zero. Any one of the elementary quantities of work represented by the type-symbol $I dp$ is exactly equal in amount, but of a contrary sign, to the aggregate quantity of work of all the other forces represented by $\Sigma I' dp'$. Such a system of forces is said to be in equilibrio, and the masses in equilibrium. If the latter be in motion, this motion must be uniform. Regarding the intensities of forces as always positive, the sign of the products Idp depends on the sign of dp. The sign of dp is taken *positive* when it falls on the action-line of the force, and *negative* when it falls on the action-line produced (Fig. 39).

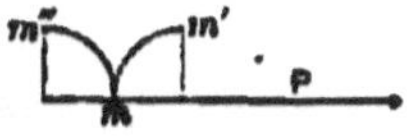

FIG. 39.

The elementary paths whose projections are dp are called *virtual velocities*. Being the actual paths described in the time dt, they have the same ratio to each other as the velocities of the points of application at the instant considered.

The products Idp, $I'dp'$, etc., are called *virtual moments*; they are the elementary quantities of work done by the forces while their points of application move over the distances whose projections on the action-lines are dp, dp', etc.

Equation (S) is the form taken by the fundamental equation in Statics, and is the mathematical statement of the principle of virtual velocities; that is, *when any system of forces is in equilibrio the algebraic sum of their virtual moments is equal to zero*. In such a system the potential energy is constant, none being transformed into kinetic energy.

The converse of this principle is also true; that is, when the algebraic sum of the virtual moments of any system of forces is equal to zero the forces are in equilibrio.

69. *Equation* (E) *referred to Rectangular Co-ordinate Axes.*—Let a, b, c be the angles made by the virtual velocity of the point of application of a force with the axes, and d the angle between this virtual velocity and the action-line of the force. Then we have

$$\cos d = \cos \alpha \cos a + \cos \beta \cos b + \cos \gamma \cos c; \quad (100)$$

and multiplying by ds,

$$ds \cos d = \cos \alpha \, ds \cos a + \cos \beta \, ds \cos b + \cos \gamma \, ds \cos c, \quad (101)$$

or

$$dp = \cos \alpha \, dx + \cos \beta \, dy + \cos \gamma \, dz. \quad \ldots \quad (102)$$

Multiplying both members of this equation by the intensity of the force, we have

$$I dp = I \cos \alpha\, dx + I \cos \beta\, dy + I \cos \gamma\, dz. \quad . \quad . \quad (103)$$

That is, *the virtual moment of any force is equal to the sum of the virtual moments of its rectangular components.*

For the whole system we have

$$\Sigma I dp = \Sigma I \cos \alpha\, dx + \Sigma I \cos \beta\, dy + \Sigma I \cos \gamma\, dz. \quad (104)$$

70. Let a', b', c' be the angles made by the elementary path of any molecule with the axes, and we have

$$1 = \cos^2 a' + \cos^2 b' + \cos^2 c', \quad . \quad . \quad . \quad (105)$$

and, multiplying by $m \dfrac{d^2 s}{dt^2} ds$,

$$m \frac{d^2 s}{dt^2} ds = m \frac{d^2 s}{dt^2} ds \cos^2 a' + m \frac{d^2 s}{dt^2} ds \cos^2 b' + m \frac{d^2 s}{dt^2} ds \cos^2 c'. \quad (106)$$

But

$$ds \cos a' = dx; \quad ds \cos b' = dy; \quad ds \cos c' = dz; \\ d^2 s \cos a' = d^2 x; \quad d^2 s \cos b' = d^2 y; \quad d^2 s \cos c' = d^2 z. \quad \Big\} (107)$$

Hence

$$m \frac{d^2 s}{dt^2} ds = m \frac{d^2 x}{dt^2} dx + m \frac{d^2 y}{dt^2} dy + m \frac{d^2 z}{dt^2} dz \quad . \quad . \quad (108)$$

That is, *the increment of the kinetic energy of any molecule is equal to the sum of the increments estimated in any three rectangular directions.*

For the whole system we have

$$\Sigma m \frac{d^2s}{dt^2}ds = \Sigma m \frac{d^2x}{dt^2}dx + \Sigma m \frac{d^2y}{dt^2}dy + \Sigma m \frac{d^2z}{dt^2}dz. \quad (109)$$

Substituting in Eq. (E), we have

$$\Sigma I \cos \alpha dx + \Sigma I \cos \beta dy + \Sigma I \cos \gamma dz =$$
$$\Sigma m \frac{d^2x}{dt^2}dx + \Sigma m \frac{d^2y}{dt^2}dy + \Sigma m \frac{d^2z}{dt^2}dz. \quad \ldots \quad (110)$$

This transformation has not in any way affected the generality of Eq. (E) which, in its new form, still embodies all the circumstances of motion of the molecules of a body, or of a system of bodies, under the action of any system of extraneous forces whatever.

71. *Application of Equation E to the Motion of a Rigid Solid.*—A *rigid solid* is a body whose molecules are supposed to preserve unchanged their *relative distances* from each other. The most general motion that can be imagined for such a hypothetical solid is one compounded of a motion of translation and rotation. Its motion of translation may be defined by that of one of its molecules, and its motion of rotation by that of the body about this molecule. In Fig. 40 let O be any fixed origin, O' the position at any instant of the particular molecule which determines the motion of translation, and m the position of any other molecule at the same instant. Let x, y, z be the co-ordinates of m referred to the fixed origin; x_0, y_0, z_0, the co-ordinates of the movable origin O' referred to the fixed, and x', y', z' the co-ordinates of m referred to the movable origin.

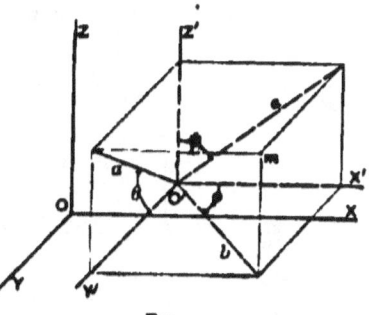

Fig. 40.

Then, supposing the axes through the movable origin O' to be always parallel to the fixed axes at O, we have

$$x = x_0 + x'; \qquad y = y_0 + y'; \qquad z = z_0 + z'; \quad . \quad (111)$$
$$dx = dx_0 + dx'; \quad dy = dy_0 + dy'; \quad dz = dz_0 + dz'. \quad . \quad (112)$$

Measuring the angles about O' as indicated in the figure to conform to Art. 35, we have, for the increments of x', y', z', due to rotation about the axes X', Y', Z',

$$X'. \begin{cases} \dfrac{dx'}{d\theta} d\theta = 0; \\ \dfrac{dy'}{d\theta} d\theta = -a d\theta \sin\theta = -z' d\theta; \\ \dfrac{dz'}{d\theta} d\theta = a d\theta \cos\theta = y' d\theta. \end{cases} \quad . \quad . \quad (113)$$

$$Y'. \begin{cases} \dfrac{dx'}{d\psi} d\psi = c d\psi \cos\psi = z' d\psi; \\ \dfrac{dy'}{d\psi} d\psi = 0; \\ \dfrac{dz'}{d\psi} d\psi = -c d\psi \sin\psi = -x' d\psi. \end{cases} \quad . \quad . \quad (114)$$

$$Z'. \begin{cases} \dfrac{dx'}{d\phi} d\phi = -b d\phi \sin\phi = -y' d\phi; \\ \dfrac{dy'}{d\phi} d\phi = b d\phi \cos\phi = x' d\phi; \\ \dfrac{dz'}{d\phi} d\phi = 0. \end{cases} \quad . \quad . \quad (115)$$

Hence we have, for the total differentials,

$$\begin{rcases} dx' = z' d\psi - y' d\phi; \\ dy' = x' d\phi - z' d\theta; \\ dz' = y' d\theta - x' d\psi. \end{rcases} \quad . \quad . \quad . \quad . \quad . \quad (116)$$

WORK AND ENERGY.

Substituting these values in Eqs. (112), we have

$$dx = dx_0 + z'd\psi - y'd\phi; \\ dy = dy_0 + x'd\phi - z'd\theta; \\ dz = dz_0 + y'd\theta - x'd\psi. \quad\quad (117)$$

Since these values apply to any point of the body, we may substitute them in Eq. (110), and we have

$$\Sigma I \cos \alpha (dx_0 + z'd\psi - y'd\phi) \\ + \Sigma I \cos \beta (dy_0 + x'd\phi - z'd\theta) \\ + \Sigma I \cos \gamma (dz_0 + y'd\theta - x'd\psi) \\ = \Sigma m \frac{d^2 x}{dt^2}(dx_0 + z'd\psi - y'd\phi) \\ + \Sigma m \frac{d^2 y}{dt^2}(dy_0 + x'd\phi - z'd\theta) \\ + \Sigma m \frac{d^2 z}{dt^2}(dz_0 + y'd\theta - x'd\psi). \quad (118)$$

But dx_0, dy_0, dz_0 relate to the movable origin, and $d\theta$, $d\psi$, $d\phi$ are independent of the position of m because the body is rigid; each of these differentials is therefore a common factor of the terms which it enters, and Eq. (118) may be written

$$\left(\Sigma I \cos \alpha - \Sigma m \frac{d^2 x}{dt^2}\right) dx_0 + \left(\Sigma I \cos \beta - \Sigma m \frac{d^2 y}{dt^2}\right) dy_0 \\ + \left(\Sigma I \cos \gamma - \Sigma m \frac{d^2 z}{dt^2}\right) dz_0 \\ + \left[\Sigma I(x' \cos \beta - y' \cos \alpha) - \Sigma m \frac{x'd^2y - y'd^2x}{dt^2}\right] d\phi \\ + \left[\Sigma I(z' \cos \alpha - x' \cos \gamma) - \Sigma m \frac{z'd^2x - x'd^2z}{dt^2}\right] d\psi \\ + \left[\Sigma I(y' \cos \gamma - z' \cos \beta) - \Sigma m \frac{y'd^2z - z'd^2y}{dt^2}\right] d\theta = 0. \quad (119)$$

72. Application of Eq. (E) to a Rigid Solid, perfectly free to Move.—The only restriction which has thus far been imposed is that Equation (119), which has been derived from Eq. (E), shall apply to a single rigid solid acted upon by any extraneous forces. The body may be subjected to any conditions whatever as to its possible motion under the action of these forces, and the values of dx_0, dy_0, dz_0, $d\theta$, $d\psi$, $d\phi$ will depend on these conditions. If no conditions be imposed, that is, if the body be *free*, then dx_0, dy_0, dz_0, $d\theta$, $d\psi$, $d\phi$, will be entirely arbitrary and independent of each other. Hence we have, by the principle of indeterminate coefficients,

$$\left.\begin{aligned} X = \Sigma I \cos \alpha = \Sigma m \frac{d^2x}{dt^2}; \\ Y = \Sigma I \cos \beta = \Sigma m \frac{d^2y}{dt^2}; \\ Z = \Sigma I \cos \gamma = \Sigma m \frac{d^2z}{dt^2}; \end{aligned}\right\} \quad \ldots \quad (T)$$

$$\left.\begin{aligned} \Sigma I(x' \cos \beta - y' \cos \alpha) = \Sigma m \frac{x'd^2y - y'd^2x}{dt^2}; \\ \Sigma I(z' \cos \alpha - x' \cos \gamma) = \Sigma m \frac{z'd^2x - x'd^2z}{dt^2}; \\ \Sigma I(y' \cos \gamma - z' \cos \beta) = \Sigma m \frac{y'd^2z - z'd^2y}{dt^2}. \end{aligned}\right\} \quad \ldots \quad (R)$$

73. Interpretation of Equations (T) and (R).—These six conditions, applicable to the case of a free rigid solid, having been derived from the general equation of energy, embody all the circumstances of motion of its molecular masses, caused by the action of extraneous forces. Considering Eqs. (T), we see that their middle members are the sums of the component intensities of the extraneous forces in the directions of the rectangular axes, and the last members are the sums of the products of the molecular masses of the body by their accelerations in the corresponding directions. But these products are the type-symbols of the intensities of extraneous forces acting on the molecular masses. Hence in a free rigid solid any system of extraneous forces may be replaced by an equal set whose points of application are the molecules of the body, and the circumstances of the motion of translation of the latter will not be changed. We see also that whether there be one or many extraneous forces acting on the solid, the connections which unite its molecules together cause the effect of these forces to be distributed throughout the whole body. Eqs. (T) therefore refer to *motion of translation, and express the fact that the algebraic sum of the component intensities of the extraneous forces, estimated in any direction, is measured by the sum of the products of the mass of each molecule by its acceleration in that direction.*

Referring now to the second members of (R), and considering the first of these, we see that it is the summation of terms of the form of

$$m'z'\frac{d^2y}{dt^2} - m'y'\frac{d^2x}{dt^2}. \quad . \quad . \quad . \quad . \quad . \quad (120)$$

But $m'\frac{d^2y}{dt^2}$ is the measure of the intensity of the force which for the instant dt acts on m' in the direction of the axis y', and x' being the co-ordinate of m' referred to O', the product $m'x'\frac{d^2y}{dt^2}$ is the moment of that force with respect to the axis z'. Similarly $m'y'\frac{d^2x}{dt^2}$ is the moment of the force $m'\frac{d^2x}{dt^2}$ with respect to the same axis z', and their difference is the resultant moment of the force acting on m' at that instant with respect to the axis z'. We see, therefore, that each molecule may be regarded as being subjected to a force of certain intensity, and *the algebraic sum of the moments of these forces at any instant, with respect to any axis, is exactly equal to that of the extraneous forces at the same instant with respect to the same axis;* and this is what is expressed by Eqs. (R).

74. If the solid be not free, the conditions to be satisfied are less than six in number. For example, if *one point* of the body be assumed as fixed, this point may be taken as the origin O', and we shall have

$$dx_0 = dy_0 = dz_0 = 0.$$

The first three terms of Eq. (119) then reduce to zero; and since $d\theta$, $d\psi$, $d\phi$ are still arbitrary and independent, this equation is satisfied when the three conditions (R) are satisfied.

If *two points* be fixed, the right line joining them will be a fixed axis, and may be taken as the axis Y'. We shall then have $dx_0 = dy_0 = dz_0 = 0$, and $d\phi = d\theta = 0$; and since $d\psi$ is not necessarily zero, Eq. (119) is satisfied when the second of Eqs. (R) is satisfied; that is, the single condition of rotation about the fixed axis; and similarly for other conditions of constraint.

75. If the molecules of the solid be in uniform motion or at rest, the forces and moments are balanced, and the body is in equilibrium both as to translation and rotation. Eqs. (T) and (R) then become

$$\left. \begin{array}{l} X = \Sigma I \cos \alpha = 0; \\ Y = \Sigma I \cos \beta = 0; \\ Z = \Sigma I \cos \gamma = 0; \end{array} \right\} \quad \ldots \ldots \quad (T')$$

$$\left. \begin{array}{l} Yx - Xy = \Sigma I(x' \cos \beta - y' \cos \alpha) = 0; \\ Xz - Zx = \Sigma I(z' \cos \alpha - x' \cos \gamma) = 0; \\ Zy - Yz = \Sigma I(y' \cos \gamma - z' \cos \beta) = 0; \end{array} \right\} \quad \ldots \quad (R')$$

which are the six conditions of equilibrium.

76. *Analytical Mechanics* consists essentially in the application of Equations (E), (T), (R), (T') and (R') to conservative systems of masses and forces. The object of the discussion is to ascertain the position of any and all molecules at any time, the nature and direction of their motions, and the configuration of the bodies of which they are the elements. The theory of the investigation is simple, but the practical application is limited by our mathematical knowledge and skill.

77. Equations (T) and (R) referred to the Centre of Mass as a Movable Origin.—Since there are as many terms in the last members of Eqs. (T) and (R) as there are molecules in the body, these equations are not in a convenient form for discussion. By taking the movable origin at the centre of mass, the resulting equations are no less general than before, while the solution of practical problems is much simplified.

78. Equations of Translation.—In the first of Eqs. (T),

$$X = \Sigma I \cos \alpha = \Sigma m \frac{d^2 x}{dt^2}, \quad \ldots \quad (121)$$

substitute for d^2x its value obtained by differentiating Eq. (112), and we have

$$X = \Sigma I \cos \alpha = \Sigma m \frac{d^2 x}{dt^2} = \Sigma m \frac{d^2 x_0}{dt^2} + \Sigma m \frac{d^2 x'}{dt^2}. \quad (122)$$

But d^2x_0 is a common factor in the term which it enters, and from the principle of the centre of mass, Eqs. (64), we have

$$\Sigma m d^2 x' = M d^2 \bar{x}, \quad \ldots \quad (123)$$

and Eq. (122) becomes

$$X = \Sigma I \cos \alpha = M \frac{d^2 x_0}{dt^2} + M \frac{d^2 \bar{x}}{dt^2}. \quad \ldots \quad (124)$$

Taking the movable origin at the centre of mass, we have $\bar{x} = 0$, and Eqs. (T) become

$$\left. \begin{array}{l} X = \Sigma I \cos \alpha = M \dfrac{d^2 x_0}{dt^2}; \\ Y = \Sigma I \cos \beta = M \dfrac{d^2 y_0}{dt^2}; \\ Z = \Sigma I \cos \gamma = M \dfrac{d^2 z_0}{dt^2}; \end{array} \right\} \quad \ldots \quad (T_m)$$

from which the motion of translation of the centre of mass of a free rigid solid, under the action of *incessant* forces, can be found.

If the forces be *impulsions*, then Eqs. (T) become

$$X = \Sigma I_t \cos \alpha = \Sigma m \frac{dx}{dt};$$

$$Y = \Sigma I_t \cos \beta = \Sigma m \frac{dy}{dt};$$

$$Z = \Sigma I_t \cos \gamma = \Sigma m \frac{dz}{dt};$$

which, when the movable origin is the centre of mass, become

$$\left.\begin{array}{l} X = \Sigma I_t \cos \alpha = M \dfrac{dx_0}{dt} = MV_x; \\[4pt] Y = \Sigma I_t \cos \beta = M \dfrac{dy_0}{dt} = MV_y; \\[4pt] Z = \Sigma I_t \cos \gamma = M \dfrac{dz_0}{dt} = MV_z; \end{array}\right\} \quad \ldots \quad (\mathrm{T_m'})$$

from which the motion of translation of the centre of mass of a free rigid solid, under the action of impulsions, can be found.

From Eqs. ($\mathrm{T_m}$) and ($\mathrm{T_m'}$) we draw the following conclusions:

(1) *That the motion of the centre of mass of a free rigid solid, under the action of extraneous forces, is entirely independent of the relative positions of the molecular masses, since their co-ordinates have disappeared from the equations of motion.*

(2) *That the motion of the centre of mass depends only upon the mass of the body and the intensities and directions of the extraneous forces, and is independent of the points of application of the forces.*

(3) *That the motion of the centre of mass will be precisely the same as that of a material point whose mass is equal to that of the body, sub-*

jected to the action of forces equal to the given forces in intensity and having the same direction.

79. To illustrate these conclusions, let us consider the motion of translation of the centre of mass of the free rigid solid A, Fig. 41. first, under the action of the several incessant forces P', P'', P''' and P^{iv}, and, second, under the action of the impulsions $P_{\prime}$, $P_{\prime\prime}$, $P_{\prime\prime\prime}$ and P_{iv}.

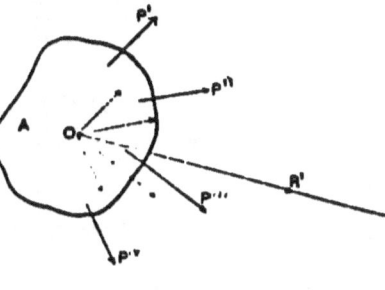

Fig. 41.

Let O be the centre of mass, and let forces equal to the extraneous forces in intensity and direction be supposed applied to it. Let R' be the resultant of those forces which have O as their common point of application; then from the above principles it can be asserted that under the action of the given incessant forces the centre of mass will move along the right line OR' with a *constant acceleration* equal to $\frac{R'}{M}$; and if the forces be impulsions, that the centre of mass will move along the right line $OR_{\prime}$ with a *constant velocity* equal to $\frac{R_{\prime}}{M}$.

80. *Equations of Rotation.*—When *the forces are incessant* and the movable origin is the centre of mass, Eqs. (R) readily reduce to

$$\left. \begin{array}{l} L = Yx - Xy = \Sigma l(x'\cos\beta - y'\cos\alpha) = \Sigma m \dfrac{x'd^2y' - y'd^2x'}{dt^2}; \\[4pt] M = Xz - Zx = \Sigma l(z'\cos\alpha - x'\cos\gamma) = \Sigma m \dfrac{z'd^2x' - x'd^2z'}{dt^2}; \\[4pt] N = Zy - Yz = \Sigma l(y'\cos\gamma - z'\cos\beta) = \Sigma m \dfrac{y'd^2z' - z'd^2y'}{dt^2}. \end{array} \right\} (R_n)$$

To show this, let us reduce the second member of the first of Eqs. (R); we have

$$\Sigma m \frac{x'd^2y - y'd^2x}{dt^2} = \Sigma m \frac{x'd^2y_0 + x'd^2y' - y'd^2x_0 - y'd^2x'}{dt^2}$$

$$= \frac{d^2y_0 \Sigma mx' - d^2x_0 \Sigma my' + \Sigma mx'd^2y' - \Sigma my'd^2x'}{dt^2}$$

$$= \Sigma m \frac{x'd^2y' - y'd^2x'}{dt^2},$$

since, by the principle of the centre of mass, $\Sigma mx' = \Sigma my' = 0$.
Similarly, *when the forces are impulsions*, Eqs. (R) become

$$\left. \begin{aligned} L_{,} &= Yx - Xy = \Sigma I_{,}(x' \cos \beta - y' \cos \alpha) = \Sigma m \frac{x'dy - y'dx}{dt}; \\ M_{,} &= Xz - Zx = \Sigma I_{,}(z' \cos \alpha - x' \cos \gamma) = \Sigma m \frac{z'dx - x'dz}{dt}; \\ N_{,} &= Zy - Yz = \Sigma I_{,}(y' \cos \gamma - z' \cos \beta) = \Sigma m \frac{y'dz - z'dy}{dt}; \end{aligned} \right\} (125)$$

which, when referred to the centre of mass as a centre of rotation, become

$$\left. \begin{aligned} L_{,} &= Yx - Xy = \Sigma I_{,}(x' \cos \beta - y' \cos \alpha) = \Sigma m \frac{x'dy' - y'dx'}{dt}; \\ M_{,} &= Xz - Zx = \Sigma I_{,}(z' \cos \alpha - x' \cos \gamma) = \Sigma m \frac{z'dx' - x'dz'}{dt}; \\ N_{,} &= Zy - Yz = \Sigma I_{,}(y' \cos \gamma - z' \cos \beta) = \Sigma m \frac{y'dz' - z'dy'}{dt}. \end{aligned} \right\} (R_m')$$

81. If one point of the body be fixed, we have, by taking it as the origin O',

$$d^2x = d^2x'; \quad d^2y = d^2y'; \quad d^2z = d^2z';$$

and Eqs. (R) reduce to the form of Eqs. (R_m) or (R_m'), independently of the principle of the centre of mass.

Also, if an axis be fixed, that one of Eqs. (R) which applies to the axis in question reduces to the corresponding one of Eqs. (R_m) or (R_m'), according as the forces are incessant or impulsive.

82. In Eqs (R_m) and (R_m') the co-ordinates of the centre of mass do not appear; therefore the motion of rotation of the body about the centre of mass is independent of the position of this centre, and will be the same whether it be considered at rest or in the state of its actual motion. This exhibits the complete independence of the motions of translation and of rotation, and permits the investigation of either as if the other did not exist. By means of Eqs. (T_m) the position, velocity and acceleration of the centre of mass can be theoretically determined at any time, and by Eqs. (R_m) the corresponding positions, angular velocities and accelerations of every molecule with respect to the centre of mass; and thus the configuration of the whole body about this point can be determined for the same instant. The mathematical difficulties, however, due to integration, limit their application to but a few simple cases. It is to be noted that when the problem involves incessant forces, Eqs. (T_m) and (R_m), and when impulsions alone, Eqs. (T_m') and (R_m'), are to be used.

General Theorem of Energy applied to a Free Rigid Body whose Centre of Mass is referred to a Fixed Point in Space.

83. *Translation under Incessant Forces.*—Multiply Eqs. (T_m) by dx, dy, dz, respectively, add the results and integrate, and we have

$$\int (Xdx + Ydy + Zdz) = M \int \frac{dxd'x + dyd'y + dzd'z}{dt'}$$

$$= \frac{M}{2}\left(\frac{dx^2 + dy^2 + dz^2}{dt^2}\right) + C = \frac{MV^2}{2} + C, \quad (126)$$

in which x, y, z are the co-ordinates of the centre of mass referred to the fixed origin, V is the variable velocity with respect to the same point, and M is the mass of the body. The first member,

$$\int (Xdx + Ydy + Zdz), \quad \ldots \quad (127)$$

is, Art. 66, an expression for the quantity of work done by the extraneous forces, or the total quantity of potential energy expended by them, and whose particular value in any case will be determined when the limits of the integration are fixed. The constant C of the second member is evidently the kinetic energy which existed in the body at the instant the extraneous forces began to act; for at that instant, say t_1, their work is zero, and if V_1 be the corresponding velocity, then

$$C = -\frac{MV_1^2}{2}. \quad \ldots \quad (128)$$

The first term of the second member, $\frac{MV^2}{2}$, is the total kinetic energy at any time, and hence the whole second member,

$$\frac{MV_2^2 - MV_1^2}{2}, \quad \ldots \quad (129)$$

being the difference between that possessed by the body at any time and that when the extraneous forces began to act on the body, is the exact equivalent of the potential energy represented by the first member. Here we have, as should have been expected, the general law of the transformation of energy.

84. The first member may be integrated when the component forces in the directions of the co-ordinate axes are constant, and when, if these forces be variable, it becomes a known differential function of the three co-ordinates x, y, z. In the first case, let R be the constant intensity of the resultant of the forces, and a, b, c the angles which its action-line makes with the co-ordinate axes. Then we have between the limits 1 and 2

$$R(x \cos a + y \cos b + z \cos c)_1^2 = F'(x,y,z)_1^2 = M \frac{(V_2^2 - V_1^2)}{2}. \quad (130)$$

In the second case, in order that integration may be possible, the intensities of the forces must be functions of x, y, z, and we have

$$F''(x, y, z),' = M \frac{V_2^2 - V_1^2}{2}. \quad \ldots \quad (131)$$

Hence we may write Eq. (126) in either case

$$F(x, y, z),' = F(x_2, y_2, z_2) - F(x_1, y_1, z_1) = M \frac{V_2^2 - V_1^2}{2}, \quad (132)$$

as the general law of energy when forces act on a free rigid solid to give it motion of translation, under the conditions imposed above. From this equation we conclude that the velocity generated in a free rigid solid by *constant forces*, or by *variable forces whose intensities are functions of the co-ordinates of the centre of mass*, varies only with the values of the co-ordinates; and that, should the centre of mass ever return to the same position in space, its velocity will be the same as before, whether the path by which it reaches this point be the same or not.

If the forces be in equilibrio, then, Art. 68,

$$Xdx + Ydy + Zdz = 0, \quad \ldots \quad (133)$$

and we have

$$\frac{MV^2}{2} = \text{a constant}; \quad \ldots \quad (134)$$

that is, the velocity is constant.

85. *Rotation under Incessant Forces.*—If we multiply Eqs. (R_m) by $d\phi$, $d\psi$ and $d\theta$, respectively, add the results and reduce by Eqs. (116), we obtain

$$\Sigma I \cos \alpha dx' + \Sigma I \cos \beta dy' + \Sigma I \cos \gamma dz'$$
$$= \Sigma m \left(\frac{dx' d^2x' + dy' d^2y' + dz' d^2z'}{dt^2} \right), \quad (135)$$

which, by integration, becomes

$$\int \Sigma I dp = \int R dr = \tfrac{1}{2}\Sigma m\left(\frac{dx'^2+dy'^2+dz'^2}{dt^2}\right)+C' = \tfrac{1}{2}\Sigma mv^2+C', \quad (136)$$

when, in Eq. (104), we replace the sum of the virtual moments of the extraneous forces, $\Sigma I dp$, by the virtual moment of the resultant, $R dr$. But dr, being the elementary path of the point of application of the resultant projected on the action-line of the resultant, is equal to the product of the path described by a point at a unit's distance from the centre of mass by the lever arm of R. Let k be this lever arm, and ds the elementary arc at a unit's distance, and we have

$$\int R k\, ds = \tfrac{1}{2}\Sigma mv^2 + C'. \quad \ldots \quad (137)$$

The first member is the general expression for the work of rotation done by the resultant of the system, or the potential energy transformed; $\tfrac{1}{2}\Sigma mv^2$ is the kinetic energy of rotation of the body with respect to the centre of mass; C' is the kinetic energy of rotation in the body before any has been transferred to it by the extraneous forces, and is equal to $-\tfrac{1}{2}\Sigma mv_0^2$; therefore we have

$$\int R k\, ds = \tfrac{1}{2}\Sigma mv^2 - \tfrac{1}{2}\Sigma mv_0^2 \quad \ldots \quad (138)$$

for the total kinetic energy of rotation put into the body by the expenditure of the equivalent amount of potential energy $\int R k\, ds$.

86. Adding together Eqs. (132) and (138), we have

$$\int_0^1 (X dx + Y dy + Z dz) + \int_0^1 R k\, ds$$
$$= M\frac{V_1^2 - V_0^2}{2} + \tfrac{1}{2}\Sigma m(v_1^2 - v_0^2). \quad (139)$$

in which the first member expresses the total expenditure of potential energy of the extraneous forces in producing motion of translation and of rotation, and the second member the equivalent quantity of kinetic energy which has resulted therefrom. If the action-line of the resultant of the extraneous forces pass through the centre of mass, the second term of the first member and the second term of the second member become zero.

Thus we see that the *point of application* is of importance in determining the effect of a force, since by supposing it to be changeable the resulting kinetic energy of rotation imparted to the body will be correspondingly varied.

87. *Translation under Impulsive Forces*.—Squaring Eqs. (T_m') and adding, we have

$$X'^2 + Y'^2 + Z'^2 = R_t'^2 = M'^2(V_x'^2 + V_y'^2 + V_z'^2) = M'^2 V'^2, \quad (140)$$

whence

$$R_t = MV. \quad \quad \quad \quad (141)$$

Hence, when a free rigid solid has been subjected only to a system of impulsions, its centre of mass will move with a constant velocity, $\dfrac{R_t}{M}$.

88. *Motion of Translation*.—In the discussion of the motion of translation of bodies two classes of problems arise. In the first, which are called *direct*, we have given the mass of the body and the forces acting upon it; and it is required to find the path of the centre of mass and all the circumstances of its motion. In those of the second class, which are called *inverse problems*, the path of the centre of mass is given, and it is required to find the forces which will cause the body to follow that path.

89. *The Direct Problem*.—To solve the direct problem we substitute in Eqs. (T_m) the mass of the body and the component intensities of the forces, and obtain

$$\frac{d^2x}{dt^2} = \frac{X}{M}; \quad \frac{d^2y}{dt^2} = \frac{Y}{M}; \quad \frac{d^2z}{dt^2} = \frac{Z}{M}; \quad \ldots \quad (142)$$

in which the accelerations are constant or variable according as the forces are constant or variable. Integrating, we have

$$\left. \begin{array}{l} \dfrac{dx}{dt} = \displaystyle\int \dfrac{X}{M}dt + C; \\ \dfrac{dy}{dt} = \displaystyle\int \dfrac{Y}{M}dt + C'; \\ \dfrac{dz}{dt} = \displaystyle\int \dfrac{Z}{M}dt + C''. \end{array} \right\} \quad \ldots \quad (143)$$

The arbitrary constants in these equations are the values of the component velocities when $t = 0$; that is, at the epoch or instant from which t is estimated.

Integrating again, we have

$$\left. \begin{array}{l} x = \displaystyle\int \left[\int \dfrac{X}{M} dt \right] dt + Ct + D; \\ y = \displaystyle\int \left[\int \dfrac{Y}{M} dt \right] dt + C't + D'; \\ z = \displaystyle\int \left[\int \dfrac{Z}{M} dt \right] dt + C''t + D''; \end{array} \right\} \quad \ldots \quad (144)$$

D, D' and D'' being the co-ordinates of the centre of mass at the epoch.

The values of C, C', C'', D, D' and D'' are called *the initial conditions*, since they are the component velocities and the co-ordinates of the centre of mass, when the forces began their action.

The integrals in Eqs. (144) can readily be found when the given forces X, Y, Z are constants, or, if variable, when they can be expressed in such terms of t as to make them known integrable expressions. Then by eliminating t from Eqs. (144) we obtain two equations containing x, y, z, and constants, which are the equations of the path. The problem is then completely solved, since we have the position, velocity and acceleration at any time, and the entire path of the body.

90. *The Inverse Problem.*—Since the centre of mass may describe the same path under different conditions of velocity and acceleration, the inverse problem is indeterminate. It may, however, be made determinate by assuming the initial conditions and *one component velocity* or *one component acceleration*.

Let the equations of the path be

$$\left.\begin{aligned}f(x, y) &= 0;\\ f(x, z) &= 0.\end{aligned}\right\} \quad \ldots \ldots \quad (145)$$

By differentiating and dividing by dt, we obtain two equations involving three component velocities, and by a second differentiation and division we get two equations containing three component accelerations. To obtain an equation connecting the velocity with the component accelerations, differentiate Eq. (126) and divide by the differential of one of the variables, as dx. We thus have

$$\frac{1}{2}M\frac{d(V^2)}{dx} = X + Y\frac{dy}{dx} + Z\frac{dz}{dx}, \quad \ldots \quad (146)$$

or

$$\frac{1}{2}\frac{d(V^2)}{dx} = \frac{d^2x}{dt^2} + \frac{d^2y}{dt^2}\cdot\frac{dy}{dx} + \frac{d^2z}{dt^2}\cdot\frac{dz}{dx}. \quad \ldots \quad (147)$$

Now if one of the component velocities be assumed, we may obtain the value of V from the equations obtained by the first differentiation of the equations of the path, since

$$V^2 = \left(\frac{dx}{dt}\right)^2 + \left(\frac{dy}{dt}\right)^2 + \left(\frac{dz}{dt}\right)^2.$$

Then we have three equations in which the three component accelerations are the only unknown quantities, and the problem may be completely solved.

If one component acceleration be assumed, we may find V from Eq. (147) by integration, and this value, together with the equations involving the component velocities, makes the solution possible. Thus, in either case the problem is determinate, and all the circumstances of the motion may be found as in the direct problem.

It is also evident that the problem may be solved by assuming *any new condition* connecting the six unknown quantities, since we already have five equations containing them.

91. *Examples of the Direct Problem.*

(1) *Constant Forces.*—Integrating Eqs. (142) twice, we have

$$\frac{dx}{dt} = \frac{X}{M}t + C; \quad \frac{dy}{dt} = \frac{Y}{M}t + C'; \quad \frac{dz}{dt} = \frac{Z}{M}t + C''; \quad (148)$$

$$\left.\begin{array}{l} x = \dfrac{X}{M}\dfrac{t^2}{2} + Ct + D; \\ y = \dfrac{Y}{M}\dfrac{t^2}{2} + C't + D'; \\ z = \dfrac{Z}{M}\dfrac{t^2}{2} + C''t + D''. \end{array}\right\} \quad \ldots \quad (149)$$

Let us suppose that the centre of mass at the epoch is at rest at the origin of co-ordinates; then Eqs. (148) and (149) become

$$\frac{dx}{dt} = \frac{X}{M}t; \quad \frac{dy}{dt} = \frac{Y}{M}t; \quad \frac{dz}{dt} = \frac{Z}{M}t; \quad \ldots \quad (150)$$

$$x = \frac{X}{M}\frac{t^2}{2}; \quad y = \frac{Y}{M}\frac{t^2}{2}; \quad z = \frac{Z}{M}\frac{t^2}{2}. \quad \ldots \quad (151)$$

Eliminating from either pair of equations pertaining to the

same axis, as x for example, the factor $\frac{X}{M}$, and indicating the velocity in that direction by the subscript, we have

$$x = \frac{V_x t}{2}. \quad \ldots \ldots \quad (152)$$

Laws of Constant Forces.—Eqs. (150), (151) and (152) express the laws of constant forces, which are:

1st. *The velocity of the centre of mass in any direction varies directly with the time, and at any instant is equal to the product of the acceleration in that direction by the time.*

2d. *The space passed over in any direction varies directly as the square of the time, and at any instant is equal to the acceleration in that direction multiplied by half the square of the time.*

3d. *The space described in any direction is equal to the component velocity in that direction at the time considered, multiplied by half the time since the epoch ; hence the space described in the first unit of time is equal to half the acceleration.*

92. (2) *Motion due to Gravity.*—Let the weight of the body, which is the only force acting, be supposed constant. Take the axis of z vertical and positive downward, and let α, β, γ be the angles which the weight Mg makes with the axes x, y, z, respectively. Then Eqs. (142) become

$$\left. \begin{array}{l} \frac{d^2 x}{dt^2} = \frac{Mg \cos \alpha}{M} = g \cos \alpha = 0; \\ \frac{d^2 y}{dt^2} = \frac{Mg \cos \beta}{M} = g \cos \beta = 0; \\ \frac{d^2 z}{dt^2} = \frac{Mg \cos \gamma}{M} = g \cos \gamma = g. \end{array} \right\} \quad \ldots \quad (153)$$

Under the general supposition that the centre of mass at the epoch is in motion and not at the origin of co-ordinates, these equations, by integration, give

$$\frac{dx}{dt} = C, \quad \frac{dy}{dt} = C', \quad \frac{dz}{dt} = gt + C'', \quad \ldots \quad (154)$$

and

$$x = Ct + D, \quad y = C't + D', \quad z = \tfrac{1}{2}gt^2 + C''t + D''. \quad (155)$$

Hence, Eqs. (153), the accelerations in the directions of x and y are zero, and if there be any motion in a horizontal direction it must be uniform; this is also shown by the first two of Eqs. (154). The acceleration g in the vertical direction is that due to gravity; and the velocity in this direction must increase algebraically, as shown by the last of Eqs. (154). From Eqs. (155) we see that the distances passed over in the directions of x and y vary directly with the time; also that the distance of the centre of mass from the origin, estimated in the vertical direction, is composed of three parts, viz., the initial co-ordinate D'', the space due to the initial velocity C'', and that due to the constant effect of gravity.

Confining the discussion to motion in a vertical direction, and omitting the accents from the constants, the equations become

$$\frac{d^2z}{dt^2} = g; \quad \frac{dz}{dt} = gt + C; \quad z = \tfrac{1}{2}gt^2 + Ct + D. \quad (156)$$

If the body start from rest at the origin, C and D will both be zero; and letting v represent the velocity and h the height fallen through, we have

$$v = gt; \quad h = \tfrac{1}{2}gt^2. \quad \ldots \quad (157)$$

Eliminating t, we have

$$v^2 = 2gh. \quad \ldots \quad (158)$$

This relation between v and h is of frequent use in problems of motion; h is called the *height due to the velocity v*, and v the *velocity due to the height h*; either may be found in terms of the other when g is known.

This relation may also be obtained from the general law of energy, since we have

$$wh = mgh = \frac{mv^2}{2}, \quad \ldots \ldots \quad (159)$$

or

$$v^2 = 2gh. \quad \quad (160)$$

If the body be projected vertically upward from the origin with an initial velocity C, then, to find the duration of its ascent and the height to which it will rise, we have, Eqs. (156),

$$0 = gt - C, \quad \text{or} \quad t = \frac{C}{g}, \quad \ldots \ldots \quad (161)$$

and

$$h = \tfrac{1}{2}gt^2 - Ct = -\frac{C^2}{2g}. \quad \ldots \ldots \quad (162)$$

Gravity therefore abstracts g units from the initial velocity every second until the body comes to rest at the altitude h, after which it will restore g units of velocity each second, and the body will reach the origin with its initial velocity.

From Eq. (162), as also from Eq. (158), we see that the height which the body will attain is equal to the square of the initial velocity divided by twice the acceleration due to gravity.

93. (3) *The Trajectory in Vacuo*.—The path described by the centre of mass of a body is called the trajectory; if the body be given an initial velocity, it is called a projectile; but the term trajectory is usually limited to the paths of projectiles intended to be thrown from guns by means of some explosive—generally gunpowder.

The discussion of the trajectory in vacuo limits the forces acting to the weights alone; and since for short distances on its surface the radii of the earth are sensibly parallel, the weight at all points of the trajectory may be considered as acting parallel to its direction at the origin. Let the projectile start from the origin, and take the axis z vertical and positive upward; the trajectory will lie in the plane of this axis and the initial direction of motion, since there is no force acting obliquely to this plane. If we take the axis of x in this plane, the differential equations of motion become

$$\frac{d^2z}{dt^2} = \frac{Z}{M} = \frac{Mg \cos \gamma}{M} = g \cos \gamma = -g; \quad . \quad . \quad (163)$$

$$\frac{d^2x}{dt^2} = \frac{X}{M} = \frac{Mg \cos \alpha}{M} = g \cos \alpha = 0; \quad . \quad . \quad . \quad (164)$$

since $\gamma = 180°$ and $\alpha = 90°$.

These equations give, by two integrations,

$$\frac{dx}{dt} = C; \quad \frac{dz}{dt} = -gt + C'; \quad . \quad . \quad . \quad . \quad . \quad . \quad (165)$$

$$x = Ct; \quad z = -\tfrac{1}{2}gt^2 + C't. \quad . \quad . \quad . \quad . \quad . \quad (166)$$

Let V be the initial velocity, and θ the angle which the trajectory at the origin makes with the axis of x. Then we have

$$C = V \cos \theta; \quad C' = V \sin \theta.$$

Substituting these values in Eqs. (165) and (166), we have

$$\frac{dx}{dt} = V \cos \theta; \quad \frac{dz}{dt} = -gt + V \sin \theta; \quad . \quad . \quad . \quad (167)$$

$$x = V \cos \theta \, t; \quad z = -\tfrac{1}{2}gt^2 + V \sin \theta \, t. \quad . \quad . \quad (168)$$

Eliminating t from Eqs. (168), we find the equation of the trajectory to be

$$s = x \tan \theta - \tfrac{1}{2}g \frac{x^2}{V^2 \cos^2 \theta} \quad \ldots \ldots \quad (169)$$

or

$$s = x \tan \theta - \frac{x^2}{4h \cos^2 \theta} \quad \ldots \ldots \quad (170)$$

when for V^2 we substitute $2gh$.

To find the co-ordinates of the highest point, we have

$$\frac{ds}{dx} = 0 = \tan \theta - \frac{2x}{4h \cos^2 \theta}, \quad \ldots \quad (171)$$

or

$$x = 2h \tan \theta \cos^2 \theta = h \sin 2\theta. \quad \ldots \quad (172)$$

This value substituted in Eq. (170) gives

$$s = h \sin 2\theta \tan \theta - \frac{h^2 \sin^2 2\theta}{4h \cos^2 \theta}$$
$$= 2h \sin^2 \theta - h \sin^2 \theta = h \sin^2 \theta. \quad \ldots \quad (173)$$

Transfer the origin to the highest point, without changing the directions of the axes, and we have

$$s + h \sin^2 \theta = (x + h \sin 2\theta) \tan \theta - \frac{(x + h \sin 2\theta)^2}{4h \cos^2 \theta}, \quad (174)$$

or

$$4h \cos^2 \theta s + 4h^2 \sin^2 \theta \cos^2 \theta = 4h \cos^2 \theta \tan \theta x$$
$$+ 4h^2 \cos^2 \theta \sin 2\theta \tan \theta - x^2 - 2h \sin 2\theta x - h^2 \sin^2 2\theta. \quad (175)$$

But

$$4h^2 \sin^2 \theta \cos^2 \theta = 4h^2 \cos^2 \theta \sin 2\theta \tan \theta - h^2 \sin^2 2\theta, \quad (176)$$

and

$$4h \cos^2 \theta \tan \theta = 2h \sin 2\theta, \quad \ldots \ldots \ldots \quad (177)$$

and Eq. (175) reduces to

$$x^2 = -4h \cos^2 \theta z, \quad \ldots \ldots \quad (178)$$

the equation of a parabola whose axis is the vertical through the highest point.

The *range*, which is that portion of the original axis of x between the two branches of the curve, is seen from Eq. (172) to be $2h \sin 2\theta$, and its maximum value for any given value of V is obtained when the angle of projection is 45°; and since this value is $2h$, the maximum range is equal to twice the height due to the initial velocity. The corresponding value of z is $\tfrac{1}{2}h$, or $\tfrac{1}{4}$ of the maximum range.

The *time* required to describe any portion of the curve is evidently

$$t_2 - t_1 = \frac{x_2 - x_1}{V \cos \theta}. \quad \ldots \ldots \quad (179)$$

The time from the origin of motion to the highest point is also given by the second of Eqs. (167),

$$\frac{dz}{dt} = 0 = -gt + V \sin \theta, \quad \ldots \ldots \quad (180)$$

or

$$t = \frac{V \sin \theta}{g}. \quad \ldots \ldots \ldots \quad (181)$$

The value for the *velocity* at any point, obtained from Eqs. (167), is

$$v^2 = \frac{dx^2 + dz^2}{dt^2} = V^2 + g^2 t^2 - 2Vgt \sin \theta, \quad \ldots \quad (182)$$

or, eliminating t by means of the first of Eqs. (168),

$$v = \sqrt{V^2 - 2g \tan \theta x + \frac{g^2 x^2}{V^2 \cos^2 \theta}} \quad \ldots \quad (183)$$

From the symmetry of the curve with respect to the vertical through the highest point, it is evident that

(1) The two branches are described in equal times.

(2) For points at the same height the angle of fall is the supplement of the angle of rise.

(3) For points at the same height the velocities are equal, since the horizontal velocity is constant and the vertical velocities are numerically equal for equal values of z.

In Eq. (170) substitute for $\cos \theta$ its value in terms of $\tan \theta$, and we have

$$z = \tan \theta x - \frac{x^2 + \tan^2 \theta x^2}{4h}, \quad \ldots \quad (184)$$

or

$$\tan \theta = \frac{2h \pm \sqrt{4h^2 - 4hz - x^2}}{x}. \quad \ldots \quad (185)$$

The point (x, z) can therefore be reached by one or by two angles of projection, according as the quantity under the radical sign is zero or positive, and cannot be reached when this quantity is negative.

It is evident that if the parabola whose equation is

$$4h^2 - 4hz - x^2 = 0 \quad \ldots \quad (186)$$

be revolved about its axis z, it will generate a surface which will be the locus of all points which can be reached by but a single angle of projection, and beyond which the projectile cannot be thrown by the initial velocity due to h. Any point within this surface can be reached by two angles of projection. This limiting surface is a paraboloid of revolution, whose vertex is on the axis of z at a height h from the origin, and the radius of whose circular section in the plane normal to the axis through the origin is $2h$.

94. (4) *The Trajectory in Air*.—The air resists the passage of a projectile through it, and thus abstracts a part of its kinetic energy. The resistance of the air is chiefly due (1) to the displacement of the air particles by the forward motion of the projectile, causing the excess of air pressure in front; (2) to the cohesion of the air and its friction on the surface of the projectile. The resultant of these forces, called the *resistance of the air*, therefore varies with the *velocity* and *form* of the projectile, the *nature of its motion*, and the *condition of the atmosphere*. It is a force of variable intensity, and its law of variation is not accurately known; hence the theorem of energy Eq. (126) cannot be directly applied, since X, Y, Z are unknown.

When the velocity of any particular projectile is known at certain points of its trajectory, the loss of energy between any two of these points, making due allowance for the effect of known forces, will be $\frac{1}{2}M(V_1^2 - V_2^2)$, which, being divided by the distance between the two points, will give the mean resistance. An approximate law of resistance may be obtained by taking these points sufficiently near together, and then varying the initial velocity so as to include all service velocities. The details of this method being given in the course of Ordnance and Gunnery, only a brief statement of the mechanical principles is here given.

Consider the projectile to have motion of translation only, and the acting forces to be the weight and the resistance of the air. Let W be the weight in pounds, g the acceleration due to gravity, r the acceleration due to the resistance of the air, V the velocity of the centre of mass at any point of the trajectory, and ϕ the angle which the trajectory makes with the axis of x. Since the resistance of the air acts along the tangent, the trajectory will lie in a vertical plane, and the axes of co-ordinates may therefore be assumed as in Art. 93.

The total acceleration at any point will be the resultant of two accelerations, one in the direction of the tangent to the trajectory and the other in the direction of the radius of curvature, Eq. (8), Art. 16. These are

$$\frac{dV}{dt} = -r - g \sin \phi. \quad \ldots \quad (187)$$

and

$$\frac{V^2}{\rho} = -g \cos \phi, \quad \ldots \quad (188)$$

respectively.

The component acceleration in the direction of the axis of x will be independent of *gravity*, and that along the radius of curvature will be independent of the *resistance of the air*. The component velocities along x and s are

$$\frac{dx}{dt} = V \cos \phi = u; \quad \ldots \quad (189)$$

$$\frac{ds}{dt} = V \sin \phi = u \tan \phi. \quad \ldots \quad (190)$$

The acceleration along x is evidently

$$\frac{du}{dt} = -r \cos \phi, \quad \ldots \quad (191)$$

whence we have

$$dt = -\frac{du}{r \cos \phi}; \quad \ldots \quad (192)$$

$$dx = -\frac{u\, du}{r \cos \phi}; \quad \ldots \quad (193)$$

and since

$$ds = dx \tan \phi, \quad \ldots \quad (194)$$

we have

$$dz = -\frac{u \tan \phi\, du}{r \cos \phi}. \quad \ldots \ldots \quad (195)$$

The component acceleration along the radius of curvature, when for ρ its value $\dfrac{ds}{d\phi}$ is substituted, becomes

$$\frac{V^2}{\rho} = V\frac{ds}{dt}\cdot\frac{1}{\rho} = V\frac{ds}{dt}\frac{d\phi}{ds} = V\frac{d\phi}{dt} = -g\cos\phi, \quad (196)$$

and therefore

$$d\phi = -\frac{g\cos\phi}{V}dt = \frac{g\,du}{Vr}. \quad \ldots \ldots \quad (197)$$

From Eqs. (192), (193), (195), (197), we have

$$\left.\begin{array}{l} t = \displaystyle\int -\frac{du}{r\cos\phi}; \\[6pt] x = \displaystyle\int -\frac{u\,du}{r\cos\phi}; \\[6pt] z = \displaystyle\int -\frac{u\tan\phi\,du}{r\cos\phi}; \\[6pt] \phi = \displaystyle\int \frac{g\,du}{Vr}. \end{array}\right\} \quad \ldots \ldots \quad (198)$$

These integrations depend upon the variables u, ϕ and r, and as there is no known relation connecting these quantities, the direct solution, which requires x, z, ϕ and t to be known throughout the entire trajectory, is impossible. The methods of approximation used in the solution of practical problems of ballistics will be found in the course of Ordnance and Gunnery.

Motion of Rotation.

95. *Moments of Inertia*.—When a body rotates about an axis, the velocity of any molecule is (Art. 17)

$$v = \omega r, \quad \ldots \ldots \ldots (199)$$

r being its distance from the axis. Its kinetic energy of rotation is therefore

$$\tfrac{1}{2} m v^2 = \tfrac{1}{2} m \omega^2 r^2, \quad \ldots \ldots (200)$$

and that of the whole body is

$$\tfrac{1}{2} \Sigma m v^2 = \tfrac{1}{2} \omega^2 \Sigma m r^2, \quad \ldots \ldots (201)$$

or one half the product of the square of the angular velocity by $\Sigma m r^2$. The latter is called the *moment of inertia* of the body with respect to the axis, and is *the sum of the products obtained by multiplying the mass of each molecule by the square of its distance from the axis.*

Since for a given angular velocity of the body about different axes the kinetic energy of rotation is directly proportional to $\Sigma m r^2$, *the moment of inertia of a body measures the capacity of the body to store up kinetic energy during a motion of rotation about the axis with respect to which the moment of inertia is taken.*

The angular velocity being the actual velocity at a unit's distance from the axis, we may write

$$\tfrac{1}{2} M_i \omega^2 = \tfrac{1}{2} \omega^2 \Sigma m r^2, \quad \ldots \ldots (202)$$

or

$$M_i = \Sigma m r^2. \quad \ldots \ldots \ldots (203)$$

Hence, $\Sigma m r^2$ *measures the mass which would, if concentrated at a unit's distance from the axis, have the same moment of inertia as the body with respect to that axis.*

96. Radius and Centre of Gyration.—Let M be the mass of the body, and write

$$Mk^2 = \Sigma mr^2. \quad \ldots \ldots \ldots (204)$$

Solving with respect to k, we have

$$k = \sqrt{\frac{\Sigma mr^2}{M}}. \quad \ldots \ldots (205)$$

The distance k is called a *radius of gyration*, and its extremity not on the axis, a *centre of gyration*. When the axis passes through the centre of mass the radius and centre of gyration are called *principal*, and such a radius is generally denoted by k_1.

From Eq. (204) we see that *if the whole mass of the body be concentrated at the centre of gyration the moment of inertia of the body with respect to the axis will not be changed.* The radius of gyration may therefore be defined to be *the distance from the axis at which the whole mass of the body may be concentrated without changing its moment of inertia.*

Since the kinetic energy of rotation of a body depends only on the angular velocity and moment of inertia, we see that for a given value of ω the kinetic energy of rotation is the same as if a mass equal to Σmr^2 were concentrated at a unit's distance from the axis, or the whole mass of the body at the distance k from the axis.

97. The Momental Ellipsoid.—Let it be required to find the relations existing between the moments of inertia of a body with respect to all right lines passing through a single point. Let the assumed point be taken as the origin (Fig. 42), and let α, β, γ be the type-symbols of the angles made by the right lines with the co-ordinate axes. Let m be the mass of one of the molecules of the body, and we have for its moment of inertia, with respect to OR,

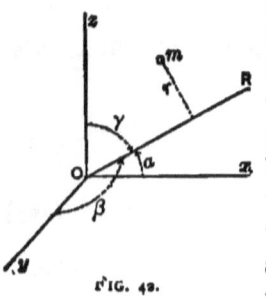

FIG. 42.

$$mr^2 = m[x^2 + y^2 + z^2 - (x \cos \alpha + y \cos \beta + z \cos \gamma)^2]. \quad (206)$$

Summing the moments of inertia for all the molecules, we have, for the moment of inertia of the body with respect to OR,

$$\Sigma mr^2 = \Sigma m[x^2 + y^2 + z^2 - (x\cos\alpha + y\cos\beta + z\cos\gamma)^2]$$
$$= \Sigma m[(x^2+y^2+z^2)(\cos^2\alpha + \cos^2\beta + \cos^2\gamma) - (x\cos\alpha + y\cos\beta + z\cos\gamma)^2]$$
$$= \Sigma m(y^2 + z^2)\cos^2\alpha + \Sigma m(x^2 + z^2)\cos^2\beta + \Sigma m(x^2 + y^2)\cos^2\gamma$$
$$- 2\Sigma myz\cos\beta\cos\gamma - 2\Sigma mxz\cos\alpha\cos\gamma - 2\Sigma mxy\cos\alpha\cos\beta. \quad (207)$$

But $\Sigma m(y^2 + z^2)$, $\Sigma m(x^2 + z^2)$ and $\Sigma m(x^2 + y^2)$ are the moments of inertia of the body with respect to the axes x, y, z, respectively. Representing these moments of inertia by A, B, C, we have

$$\Sigma mr^2 = A\cos^2\alpha + B\cos^2\beta + C\cos^2\gamma - 2\Sigma myz\cos\beta\cos\gamma$$
$$- 2\Sigma mxz\cos\alpha\cos\gamma - 2\Sigma mxy\cos\alpha\cos\beta. \quad (208)$$

Lay off on OR a distance from O equal to $\dfrac{1}{\sqrt{\Sigma mr^2}}$, and let x', y', z' be the co-ordinates of the point thus determined. Then we have

$$\left.\begin{array}{l} x' = \dfrac{\cos\alpha}{\sqrt{\Sigma mr^2}}; \\[4pt] y' = \dfrac{\cos\beta}{\sqrt{\Sigma mr^2}}; \\[4pt] z' = \dfrac{\cos\gamma}{\sqrt{\Sigma mr^2}}; \end{array}\right\} \quad (209)$$

or

$$\left.\begin{array}{l} \cos\alpha = x'\sqrt{\Sigma mr^2}; \\ \cos\beta = y'\sqrt{\Sigma mr^2}; \\ \cos\gamma = z'\sqrt{\Sigma mr^2}. \end{array}\right\} \quad (210)$$

Substituting these values of the cosines in Eq. (208), we have

$$1 = Ax'^2 + By'^2 + Cz'^2 - 2(\Sigma myz)y'z' - 2(\Sigma mxz)x'z'$$
$$- 2(\Sigma mxy)x'y'. \quad (211)$$

which is the equation of the locus of all points that are at a distance from the origin equal to the reciprocal of the square root of the moment of inertia of the body with respect to the line upon which this distance is laid off. We see from the equation that this locus is a surface of the second order; and, since the radius-vector is always finite, it is an ellipsoid.

This ellipsoid is called the *momental ellipsoid of inertia*, for the reason that the square of the reciprocal of any one of its semi-diameters is the moment of inertia of the body with respect to the coincident right line. It presents a geometrical image of the values of the moments of inertia of the body with respect to all lines radiating from the assumed point.

The greatest moment of inertia is that with respect to the shortest diameter, and the least is that with respect to the greatest diameter. As all semi-diameters of the cyclic sections are equal to the mean semi-axis of the ellipsoid, the moments of inertia with respect to these lines are equal to each other. The origin O having been assumed at pleasure, it is evident that there is a momental ellipsoid of the body for each point in space.

98. *Principal Axes.*—Since the equation of an ellipsoid when referred to its centre and axes takes the form of

$$Ax''^2 + By''^2 + Cz''^2 = 1, \quad \ldots \ldots \quad (212)$$

we see that for at least one set of rectangular co-ordinate axes through any point in space we must have the conditions

$$\left. \begin{array}{l} \Sigma mxy = 0; \\ \Sigma mxz = 0; \\ \Sigma myz = 0. \end{array} \right\} \quad \ldots \ldots \quad (213)$$

The axes of figure of the momental ellipsoid are called *principal axes* at the point considered; and since for such axes we have the conditions expressed by Eqs. (213), the latter are called *the conditions for principal axes.*

MOTION OF ROTATION.

The quantities Σmxy, Σmxz, Σmyz reduce to zero for principal axes because the sum of the positive and negative products arising from the signs of the co-ordinates x, y, z are numerically equal to each other. The moments of inertia for such axes are called *principal* moments of inertia; they evidently include the greatest and least moments of inertia at the point.

The value of Σmr^2 for any axis, Eq. (211), in terms of the principal moments of inertia at the point considered, becomes

$$\Sigma mr^2 = A \cos^2 \alpha + B \cos^2 \beta + C \cos^2 \gamma; . \quad (214)$$

that is, *the moment of inertia of any body with respect to any line whatever is equal to the sum of the products obtained by multiplying the principal moments of inertia at any point of the line, respectively, by the squares of the cosines of the angles which the line makes with the principal axes at the point.*

99. It is readily seen, Eq. (214), that when the principal moments of inertia of a body are known at any point, all its other moments of inertia with respect to that point may be determined; and it will now be shown that if the principal moments of inertia be known at the centre of mass, the moments of inertia with respect to all lines whatever can be readily computed. Let any right line be taken as the axis of z; then the moment of inertia with respect to this line is

$$\Sigma mr^2 = \Sigma m(x^2 + y^2) . . . . (215)$$

Let x_0, y_0 be the co-ordinates of the centre of mass referred to the assumed axes, and x', y' the co-ordinates of the molecules of the body referred to the centre of mass; then we have

$$x = x_0 + x'; \quad y = y_0 + y';$$

which, substituted in Eq. (215), give

$$\Sigma mr^2 = \Sigma m(x^2 + y^2) = \Sigma m[(x_0 + x')^2 + (y_0 + y')^2]$$
$$= \Sigma m(x_0^2 + y_0^2) + \Sigma m(x'^2 + y'^2) + 2x_0 \Sigma mx' + 2y_0 \Sigma my'. \quad (216)$$

Placing $x_0^2 + y_0^2 = d^2$, and remembering that by the principle of the centre of mass

$$\Sigma mx' = \Sigma my' = 0,$$

we have

$$\Sigma mr^2 = Md^2 + \Sigma m(x'^2 + y'^2). \quad \ldots \quad (217)$$

Therefore, *the moment of inertia of the body with respect to any line in space is equal to its moment of inertia with respect to a parallel line through the centre of mass, increased by the product of the mass of the body by the square of the perpendicular distance from the centre of mass to the given line.* The least principal moment of inertia at the centre of mass is therefore the least of all the moments of inertia of the body.

100. *Discussion of the Momental Ellipsoids of a Body.*—Let A, B, C be the principal moments of inertia at the centre of mass.

(1) Suppose $A = B = C$. Then the central ellipsoid is a sphere, and therefore all moments of inertia at the centre of mass are equal, and all axes through it are principal. For every other point in space the ellipsoid is a prolate spheroid whose axis passes through the centre of mass; for, the moment of inertia with respect to this line is the same as the central moments of inertia, while those with respect to all lines perpendicular to this are greater than the central moments of inertia and equal to each other; and these lines are principal axes since the moment of inertia with respect to the line through the centre of mass is the least of all the moments of inertia at the point in question.

(2) Suppose $A > B$ and $B = C$. The central ellipsoid is an oblate spheroid whose axis is that of the greatest moment of inertia. There are two points on the axis of the spheroid at which the ellipsoid is a sphere, and they are found thus: At these points all moments of inertia must be equal to A. Then, denoting by x the distance of these points from the centre, we have

$$A = B + Mx^2 = C + Mx'^2; \quad \ldots \quad (218)$$

whence

$$x = \pm\sqrt{\frac{A-B}{M}}. \quad \ldots \quad (219)$$

It is evident that the ellipsoid can be a sphere at no other point.

(3) Suppose $A = B$ and $B > C$. The central ellipsoid is a prolate spheroid whose axis is that of C. There is no point at which the ellipsoid can be a sphere.

(4) When $A > B > C$, the central ellipsoid is one of three unequal axes at the centre of mass, and cannot be a sphere at any point in space.

101. *Determination of the Moment of Inertia.*—The moment of inertia may sometimes be found by the summation of the separate values of mr^2.

Whenever the body is one whose density and boundary vary by some law of continuity, we may write

$$m = dM = \delta dV \quad \ldots \quad (220)$$

and

$$\Sigma mr^2 = \int r^2 dM = \int r^2 \delta dV; \quad \ldots \quad (221)$$

from which the moment of inertia can be found whenever the expression can be integrated between the limits that determine its volume.

Having found the moment of inertia of a body with respect to any line, that with respect to a parallel line may be found by Eq. (217); and having found the principal moments of inertia at any point, that with respect to any other line through the point can be found from Eq. (214).

In the following examples let δ represent the density, ω the area of cross-section of a material line, and t the thickness of a material surface.

Ex. 1. *A Uniform Straight Rod.*—Let the axis be perpendicular to the rod at its middle point, and represent the length of the rod by $2a$; then we have

$$dM = \delta\omega dx \quad \text{and} \quad \Sigma mr^2 = Mk_i^2 = \delta\omega \int_{-a}^{a} x^2 dx = 2a\delta\omega \frac{a^2}{3}.$$

Whence, since $M = 2a\delta\omega$, we have

$$Mk_i^2 = M\frac{a^2}{3} \quad \text{and} \quad k_i^2 = \frac{a^2}{3}. \quad \ldots \quad (222)$$

The centre of gyration is therefore at a distance from the centre of mass equal to $\dfrac{a}{1.732+} = .577a$, nearly.

For an axis perpendicular to the rod at any distance d from the centre we have

$$\Sigma mr^2 = M\left(\frac{a^2}{3} + d^2\right); \quad \ldots \quad (223)$$

which becomes, for the perpendicular axis at either extremity,

$$M\frac{4}{3}a^2.$$

Ex. 2. *A Circular Arc*, subtending an angle 2θ at the centre and whose radius is a.

1. Axis perpendicular to the plane of the arc through its centre.

$$\Sigma mr^2 = \delta\omega \int_0^{2\theta} a^2 d\theta = 2\delta\omega a^2 \theta = Ma^2;$$

and, for the whole circumference,

$$\Sigma mr^2 = 2\pi\delta\omega a^3 = Ma^2 \quad \ldots \ldots \ldots (224)$$

2. **Axis in the plane of the arc through its centre and middle point.**

$$\Sigma mr^2 = \delta\omega \int_{-\theta}^{\theta} a^3 \sin^2\theta d\theta$$
$$= \delta\omega a^3(\theta - \sin\theta\cos\theta);$$

and, for the whole circumference,

$$\Sigma mr^2 = \pi\delta\omega a^3 = M\frac{a^2}{2}. \quad \ldots \ldots \ldots (225)$$

From Eqs. (224) and (225) Σmr^2 can be found for any right line passing through the centre by the application of Eq. (214).

Ex. 3. *A Rectangular Plate* whose sides are $2a$ and $2b$. Take the centre of the plate as the origin, and the axes parallel to its sides; then for the axis x, perpendicular to $2b$, we have $dM = 2a\delta t dy$, and therefore

$$\Sigma mr_x^2 = 2a\delta t \int_{-b}^{b} y^2 dy = \frac{4}{3}\delta tab^3 = M\frac{b^2}{3}; \quad \ldots (226)$$

and similarly, for the axis y,

$$\Sigma mr_y^2 = M\frac{a^2}{3}. \quad \ldots \ldots \ldots (227)$$

For the axis z we have

$$\Sigma m(x^2 + y^2) = \Sigma m(y^2 + z^2) + \Sigma m(x^2 + z^2),$$

since $z = 0$; hence

$$\Sigma mr_z^2 = M\frac{(a^2 + b^2)}{3}. \quad \ldots \ldots \quad (228)$$

If *the plate be square*, $a = b$, and we have

$$\left.\begin{array}{l}\Sigma mr_x^2 = \Sigma mr_y^2 = M\dfrac{a^2}{3}; \\ \Sigma mr_z^2 = M\dfrac{2a^2}{3}.\end{array}\right\} \ldots \ldots \quad (229)$$

Ex. 4. *A Triangular Area about an Axis through a Vertex.*—
Let ABC, Fig. 43, be the triangle, and take the axes x and y through the vertex in its plane; let β and β' be the distances of the vertex B from y and x respectively, and γ and γ' those of the vertex C from the same axes, and let $AD = l$; then we have for the triangle ABD

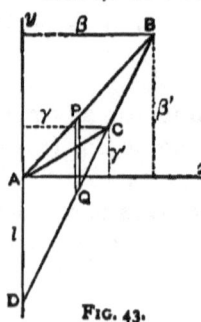

Fig. 43.

$$dM = \delta t PQ dx = \delta tl \frac{\beta - x}{\beta} dx.$$

$$\therefore \Sigma mr_y^2 = \delta tl \int_0^\beta \left(1 - \frac{x}{\beta}\right) x^2 dx = \delta tl \frac{\beta^3}{12}. \quad . \quad . \quad (230)$$

Similarly we have, for the triangle ADC,

$$\Sigma mr_y^2 = \delta tl \int_0^\gamma \left(1 - \frac{x}{\gamma}\right) x^2 dx = \delta tl \frac{\gamma^3}{12}. \quad . \quad . \quad (231)$$

But Σmr_y^2 of ABC is equal to the difference between that of ABD and ACD; therefore we have, for ABC,

$$\Sigma mr_y{}^2 = \delta u \frac{\beta^3 - \gamma^3}{12}. \quad \ldots \ldots \quad (232)$$

The mass of ABC is evidently $\delta u \dfrac{\beta - \gamma}{2}$; hence

$$\Sigma mr_y{}^2 = \frac{M}{6}(\beta^2 + \beta\gamma + \gamma^2), \quad \ldots \ldots \quad (233)$$

which is a general formula for the moment of inertia of any triangle with reference to an axis in its plane, passing through its vertex and being wholly without the triangle. Similarly we have

$$\Sigma mr_z{}^2 = \frac{M}{6}(\beta'^2 + \beta'\gamma' + \gamma'^2), \quad \ldots \quad (234)$$

and, for the axis s at A,

$$\Sigma mr_s{}^2 = \frac{M}{6}(\beta^2 + \beta\gamma + \gamma^2 + \beta'^2 + \beta'\gamma' + \gamma'^2). \quad (235)$$

Ex. 5. *A Triangular Area about any Axis whatever.*—At the middle point of each side let one third the mass of the triangle be concentrated; then the centre of gravity of the three material points coincides with that of the triangle.

The moment of inertia of the material points with reference to any line Ay drawn through A is

$$\frac{M}{3}\left\{\left(\frac{\beta+\gamma}{2}\right)^2 + \frac{\beta^2}{4} + \frac{\gamma^2}{4}\right\} = \frac{M}{6}(\beta^2 + \beta\gamma + \gamma^2), \quad (236)$$

and with reference to a line through A perpendicular to the plane of the triangle is

$$\frac{M}{3}\left\{\left(\frac{\beta^2}{4}+\frac{\beta'^2}{4}\right)+\left(\frac{\gamma^2}{4}+\frac{\gamma'^2}{4}\right)+\left(\frac{(\beta+\gamma)^2}{4}+\frac{(\beta'+\gamma')^2}{4}\right)\right\}$$
$$=\frac{M}{6}(\beta^2+\beta\gamma+\gamma^2+\beta'^2+\beta'\gamma'+\gamma'^2); \quad (237)$$

and these moments of inertia are the same as those of the triangle with respect to the same lines.

The moments of inertia expressed by Eqs. (235) and (237) are evidently the greatest moments of inertia at the point, and they are therefore principal moments of inertia; hence two of the principal axes lie in the plane of the triangle. But, Eqs. (233) and (236), the sections of the ellipsoids by this plane for the point A of the triangle and system of points coincide throughout at least half their length; therefore the principal axes in this plane are also equal and coincident, and the ellipsoids coincide throughout. Hence the ellipsoids at the common centre of mass are one and the same ellipsoid. The triangle and system of material points, having then the same central ellipsoid, have equal moments of inertia with respect to all lines in space. Therefore, to determine the moment of inertia of a triangular area, find that of three masses each equal to $\dfrac{M}{3}$ at the middle of its sides with respect to the given line, and it will be the required moment of inertia of the triangle.

Ex. 6. *An Elliptical Area.*—Let the equation of the ellipse be

$$a^2y^2+b^2x^2=a^2b^2;$$

then, for a line through its centre coincident with its major axis, we have

$$\left.\begin{aligned}\Sigma m r_x^2 &= 4\delta t\int_0^a\int_0^y y^2 dx dy \\ &= \frac{4\delta t b^3}{3a^3}\int_0^a(a^2-x^2)^{\frac{3}{2}}dx \\ &= \pi a b \delta t\frac{b^2}{4}=M\frac{b^2}{4};\end{aligned}\right\} \quad \cdots \quad (238)$$

after substituting $\frac{b}{a}(a^2 - x^2)^{\frac{1}{2}}$ for y after the first integration.

Similarly, for the minor axis, we have

$$\left.\begin{array}{l} \Sigma m r_y^2 = 4\delta t \int_0^b \int_0^x x^2 dy dx \\ = \pi a b \delta t \dfrac{a^2}{4} = M\dfrac{a^2}{4}, \end{array}\right\} \quad \ldots \quad (239)$$

and by combination we have

$$\Sigma m r_z^2 = M\frac{(a^2 + b^2)}{4}. \quad \ldots \ldots (240)$$

For an axis in the plane of the ellipse coincident with any radius vector r we have, Eq. (214), since $y^2 = r^2 \sin^2 \alpha$ and $x^2 = r^2 \cos^2 \alpha$,

$$\Sigma m r_r^2 = M\frac{(a^2 \sin^2 \alpha + b^2 \cos^2 \alpha)}{4} = M\frac{a^2 b^2}{4 r^2}. \quad (241)$$

For a *circular area* we have

$$\Sigma m r_x^2 = M\frac{a^2}{4}; \quad \Sigma m r_z^2 = M\frac{a^2}{2}.$$

Ex. 7. *An Ellipsoid.*—Let the equation of the ellipsoid be

$$\frac{x^2}{a^2} + \frac{y^2}{b^2} + \frac{z^2}{c^2} = 1;$$

then the area of any section perpendicular to the axis of x is

$$\pi \cdot \frac{c}{a}\sqrt{a^2 - x^2} \frac{b}{a}\sqrt{a^2 - x^2} = \frac{\pi c b}{a^2}(a^2 - x^2),$$

and therefore

$$dM = \delta \frac{\pi cb}{a^2}(a^2 - x^2)dx.$$

From Eq. (240) we see that the square of the radius of gyration for an ellipse with respect to the normal through its centre is equal to the sum of the squares of its semi-axes divided by 4; therefore we have

$$\left.\begin{aligned}\Sigma mr_x^2 &= \delta \int_{-a}^{a} \frac{\pi bc}{a^2}(a^2 - x^2)\frac{b^2 + c^2}{4a^2}(a^2 - x^2)dx \\ &= \delta \pi \frac{bc(b^2 + c^2)}{4a^4}\int_{-a}^{a}(a^2 - x^2)^2 dx \\ &= \frac{4}{3}\delta \pi abc\frac{b^2 + c^2}{5} \\ &= M\frac{b^2 + c^2}{5}.\end{aligned}\right\} \quad . \quad (242)$$

In the same way we readily obtain

$$\left.\begin{aligned}\Sigma mr_y^2 &= M\frac{a^2 + c^2}{5}; \\ \Sigma mr_z^2 &= M\frac{a^2 + b^2}{5}.\end{aligned}\right\} \quad \ldots \quad (243)$$

For a *spheroid* whose axis is a, and $b = c$, we have

$$\Sigma mr_x^2 = M\frac{2b^2}{5} \quad \ldots \ldots \ldots \quad (244)$$

$$\Sigma mr_y^2 = \Sigma mr_z^2 = M\frac{a^2 + b^2}{5} \quad \ldots \ldots \quad (245)$$

For a *sphere*,

$$\Sigma mr_x^2 = \Sigma mr_y^2 = \Sigma mr_z^2 = M\frac{2a^2}{5} \quad \ldots \quad (246)$$

Ex. 8. *A Rectangular Parallelopipedon.*—Assume the origin at one of the angles of the solid; let the co-ordinate axes coincide with its edges, and let a, b, c be their lengths along the axes x, y, z, respectively; then

$$\left.\begin{aligned}A &= \int_0^a \int_0^b \int_0^c \delta(y^2 + z^2) dx dy dz \\ &= \frac{\delta abc(b^2 + c^2)}{3} = M\frac{b^2 + c^2}{3}; \\ B &= \frac{\delta abc(a^2 + c^2)}{3} = M\frac{a^2 + c^2}{3}; \\ C &= \frac{\delta abc(a^2 + b^2)}{3} = M\frac{a^2 + b^2}{3};\end{aligned}\right\} \quad \ldots \quad (247)$$

and for principal axes at the centre of mass, (Art. 99),

$$A = M\frac{b^2 + c^2}{12}; \quad B = M\frac{a^2 + c^2}{12}; \quad C = M\frac{a^2 + b^2}{12}. \quad (248)$$

For the *cube*, since $a = b = c$, we have for the edges

$$A = B = C = M\frac{2a^2}{3},$$

and for principal axes at the centre of mass

$$A = B = C = M\left(\frac{2a^2}{3} - \frac{a^2}{2}\right) = M\frac{a^2}{6}. \quad \ldots \quad (249)$$

Hence the momental ellipsoid at the centre of the cube is a sphere, and all moments of inertia are equal.

It might appear that the edges are principal axes at the angle of the cube from the equality of their moments of inertia. But the line joining the centre and angle is a principal axis at the angle, since the moment of inertia with respect to it is less than that with respect to any other line; hence the principal axes at the angle are the diagonal of the cube and any pair of perpendicular right lines in the normal plane to the diagonal at the angle. The moment of inertia with respect to every right line in this plane passing through the angle of the cube is readily seen to be

$$M\left(\frac{a^2}{6} + \frac{3a^2}{4}\right) = M\frac{11a^2}{12}. \quad \ldots \ldots \quad (250)$$

102. The moments of inertia with respect to central principal axes are here tabulated for convenient reference.

Mass.	Dimensions.	X.	Y.	Z.
1. Rod or Cylinder	length $= 2a$, radius $= r$	$M\dfrac{a^2}{3}$	$M\dfrac{a^2}{3}$	$M\dfrac{r^2}{2}$
2. Circular Rim	radius $= a$	Ma^2	$M\dfrac{a^2}{2}$	$M\dfrac{a^2}{2}$
3. Rectangular Plate	sides $= a$ and b	$\dfrac{1}{4}M\dfrac{b^2}{3}$	$\dfrac{1}{4}M\dfrac{a^2}{3}$	$\dfrac{1}{4}M\dfrac{a^2+b^2}{3}$
4. Elliptical Area	semi-axes $= a$ and b	$M\dfrac{b^2}{4}$	$M\dfrac{a^2}{4}$	$M\left(\dfrac{a^2+b^2}{4}\right)$
5. Circular Area	radius $= a$	$M\dfrac{a^2}{4}$	$M\dfrac{a^2}{4}$	$M\dfrac{a^2}{2}$
6. Ellipsoid	semi-axes $= a$, b and c	$M\dfrac{b^2+c^2}{5}$	$M\dfrac{a^2+c^2}{5}$	$M\dfrac{a^2+b^2}{5}$
7. Spheroid	$b = c$	$M\dfrac{2b^2}{5}$	$M\dfrac{a^2+b^2}{5}$	$M\dfrac{a^2+b^2}{5}$
8. Sphere	radius $= a$	$M\dfrac{2a^2}{5}$	$M\dfrac{2a^2}{5}$	$M\dfrac{2a^2}{5}$
9. Rectangular Parallelopipedon	edges $= a, b, c$	$\dfrac{1}{4}M\dfrac{b^2+c^2}{3}$	$\dfrac{1}{4}M\dfrac{a^2+c^2}{3}$	$\dfrac{1}{4}M\dfrac{a^2+b^2}{3}$
10. Cube	edge $= a$	$\dfrac{1}{2}M\dfrac{a^2}{3}$	$\dfrac{1}{2}M\dfrac{a^2}{3}$	$\dfrac{1}{2}M\dfrac{a^2}{3}$

From the moments of inertia above we can readily derive the corresponding radii of gyration by dividing by the mass of the body and taking the square root of the quotient.

If the body be irregular in form and be not homogeneous, the principles of the calculus cannot be applied to find its moment of inertia. In such cases the moment of inertia can be experimentally found by means of the principles of the compound pendulum, a method which will be explained subsequently.

Instantaneous Axis.

103. Whatever may be the component angular velocities of a body when rotating about a centre, the resultant angular velocity and the axis of rotation may be found by the application of the principle of the parallelopipedon of angular velocities. When the centre of mass is taken as the centre of rotation, these are called *instantaneous angular velocity* and *instantaneous axis*. At any instant the path of each molecule of the body is in a plane perpendicular to the instantaneous axis, and all points on this axis have no motion with respect to the centre of mass.

The component velocities of any molecule with respect to the centre of mass are obtained by dividing Eqs. (116) by dt; thus we have

$$\left. \begin{array}{l} \dfrac{dx'}{dt} = z'\dfrac{d\psi}{dt} - y'\dfrac{d\phi}{dt} = z'\omega_y - y'\omega_z; \\[4pt] \dfrac{dy'}{dt} = x'\dfrac{d\phi}{dt} - z'\dfrac{d\theta}{dt} = x'\omega_z - z'\omega_x; \\[4pt] \dfrac{dz'}{dt} = y'\dfrac{d\theta}{dt} - x'\dfrac{d\psi}{dt} = y'\omega_x - x'\omega_y; \end{array} \right\} \quad . \quad (251)$$

when we substitute the symbols ω_x, ω_y, ω_z for the component angular velocities $\dfrac{d\theta}{dt}$, $\dfrac{d\psi}{dt}$, $\dfrac{d\phi}{dt}$ about the co-ordinate axes x', y', z', respectively. If in these equations we make

$$\frac{dx'}{dt} = \frac{dy'}{dt} = \frac{dz'}{dt} = 0,$$

we shall obtain the equations of the locus of all those points which are at rest with respect to the centre of mass; and, since these points lie on the instantaneous axis, we have

$$\left.\begin{array}{l} z'\omega_y - y'\omega_z = 0, \\ x'\omega_z - z'\omega_x = 0, \\ y'\omega_x - x'\omega_y = 0, \end{array}\right\} \quad \ldots \ldots \quad (252)$$

for the equations of the instantaneous axis. All molecules of the body not on this right line will have, at the assumed instant, a motion with respect to the centre of mass, and will therefore rotate about it. Since this axis passes through the centre of mass, the position of the axis in space depends only on the values of the angular velocities ω_x, ω_y, ω_z; and as these values generally remain constant only for the instant dt, the instantaneous axis describes the surface of a cone whose vertex is the centre of mass.

104. Let α, β, γ be the angles which the instantaneous axis makes with the co-ordinate axes, and ω the instantaneous angular velocity; then, by the principle of the parallelopipedon of angular velocities, we have

$$\cos \alpha = \frac{\omega_x}{\omega}, \quad \cos \beta = \frac{\omega_y}{\omega}, \quad \cos \gamma = \frac{\omega_z}{\omega}, \quad . \quad (253)$$

and

$$\omega^2 = \omega_x^2 + \omega_y^2 + \omega_z^2. \quad \ldots \ldots \quad (254)$$

Hence it is necessary to find values for the component angular velocities before the position of the axis and the resultant angular velocity of the body can be determined. The two cases to consider are (1) *rotation due to the action of incessant forces*, and (2) *rotation due to impulsions*. The latter, being the simpler case, will be discussed first.

Rotation of a Rigid Solid due to Impulsive Forces.

105. When the centre of mass is the movable origin, the motion of translation of this point and the motion of rotation of the body about it have been shown to be wholly independent of each other. Hence we may regard the centre of mass as a fixed point in space, and consider only the rotation of the body about it. The body is supposed to have been subjected to a system of impulsions whose effect is completed in a very short time, and the body is then abandoned to itself, free from the action of any extraneous force whatever. It is required to find its subsequent motion of rotation in all of its particulars.

When the moments of the several impulsions may be compounded into a single resultant moment, Rk, having the centre of mass as the centre of moments, the moment axis of R is called the *resultant* or *invariable axis*, and the plane of R and k is called the *resultant* or *invariable plane*. They cannot change their direction in space unless other forces be introduced, which is not supposed.

106. Assuming Eqs. R_m', which are here applicable,

$$\left.\begin{array}{l} \Sigma I_{\prime}(x' \cos \beta - y' \cos \alpha) = \Sigma m \dfrac{x'dy' - y'dx'}{dt} = L_{\prime\prime} \\[4pt] \Sigma I_{\prime}(z' \cos \alpha - x' \cos \gamma) = \Sigma m \dfrac{z'dx' - x'dz'}{dt} = M_{\prime\prime} \\[4pt] \Sigma I_{\prime}(y' \cos \gamma - z' \cos \beta) = \Sigma m \dfrac{y'dz' - z'dy'}{dt} = N_{\prime\prime} \end{array}\right\} (R_m')$$

and substituting the values of dx', dy', dz' given in Eqs. (251), we have

$$\left.\begin{array}{l} \Sigma m \dfrac{x'dy' - y'dx'}{dt} = \Sigma m(x'^2 + y'^2)\omega_z - \Sigma m x' z' \omega_x - \Sigma m y' z' \omega_y; \\[4pt] \Sigma m \dfrac{z'dx' - x'dz'}{dt} = \Sigma m(x'^2 + z'^2)\omega_y - \Sigma m y' z' \omega_z - \Sigma m x' y' \omega_x; \\[4pt] \Sigma m \dfrac{y'dz' - z'dy'}{dt} = \Sigma m(y'^2 + z'^2)\omega_x - \Sigma m x' y' \omega_y - \Sigma m x' z' \omega_z. \end{array}\right\} (255)$$

If the axes be principal, these equations reduce to

$$\begin{aligned}\Sigma m\frac{x'dy'-y'dx'}{dt} &= \Sigma m(x'^2+y'^2)\omega_z = C\omega_z = L_i; \\ \Sigma m\frac{z'dx'-x'dz'}{dt} &= \Sigma m(x'^2+z'^2)\omega_y = B\omega_y = M_i; \\ \Sigma m\frac{y'dz'-z'dy'}{dt} &= \Sigma m(y'^2+z'^2)\omega_x = A\omega_x = N_i. \end{aligned} \quad (256)$$

Hence

$$\omega_z = \frac{L_i}{C}; \quad \omega_y = \frac{M_i}{B}; \quad \omega_x = \frac{N_i}{A}. \quad \ldots \quad (257)$$

That is, *the angular velocity due to an impulsion about a principal axis is equal to the component moment of the impulsion divided by the moment of inertia of the body, both taken with respect to that axis.*

This principle is true also for the instantaneous axis; for if z be the instantaneous axis, we have

$$\omega_x = \omega_y = 0, \quad \text{and} \quad \omega_z = \omega.$$

And substituting these values in the first of Eqs. (255), we have

$$\Sigma m\frac{x'dy'-y'dx'}{dt} = \Sigma m(x'^2+y'^2)\omega_z = C\omega = L_i. \quad (258)$$

Since Eqs. (R_m') apply to rotation about a fixed point or a fixed axis, this principle is likewise applicable to both of these cases.

107. Let Σmr^2 be the moment of inertia with respect to the instantaneous axis, and ϕ the angle between this axis and the invariable axis. Then we have

$$\omega = \frac{Rk\cos\phi}{\Sigma mr^2}; \quad \ldots \quad \ldots \quad (259)$$

hence

$$\Sigma mv^2 = \omega^2 \Sigma mr^2 = \frac{\omega^2}{\delta^2} = Rk\omega \cos\phi, \quad \ldots \quad (260)$$

δ being that semi-diameter of the central ellipsoid which coincides with the instantaneous axis.

Squaring Eqs. (256) and adding, we have

$$A^2\omega_x^2 + B^2\omega_y^2 + C^2\omega_z^2 = L_t^2 + M_t^2 + N_t^2 = R^2k^2; \quad (261)$$

Rk being the resultant moment of the system of impulsions with respect to the centre of mass.

From Eq. (260) we conclude, since Σmv^2 is constant,

(1) The instantaneous angular velocity varies directly with the length of the semi-diameter of the ellipsoid which coincides with the instantaneous axis, or inversely as the square root of the moment of inertia with respect to this axis.

(2) The angular velocity about the invariable axis, $\omega \cos\phi$, is constant; therefore, as ϕ increases or $\cos\phi$ diminishes, ω increases; that is, as the instantaneous axis increases its inclination to the invariable axis, the instantaneous angular velocity increases.

Eqs. (260) and (261), together with that of the central ellipsoid, give the circumstances of the rotary motion of a free rigid solid under the action of impulsive forces whenever we can find the value of Rk.

108. The equation of the invariable plane is

$$\frac{N_t}{Rk}x + \frac{M_t}{Rk}y + \frac{L_t}{Rk}z = 0. \quad \ldots \quad (262)$$

Call the point in which the instantaneous axis pierces the central ellipsoid the *instantaneous pole*, and let x', y', z' be its coordinates. Then the equation of the tangent plane to the ellipsoid at the instantaneous pole is

$$Axx' + Byy' + Czz' = 1. \quad \ldots \quad (263)$$

From Eqs. (252) we have

$$\frac{\omega_x}{x'} = \frac{\omega_y}{y'} = \frac{\omega_z}{z'} = \frac{\omega}{\delta} = \epsilon, \quad \ldots \quad (264)$$

which reduces Eq. (263) to

$$A\omega_x x + B\omega_y y + C\omega_z z = \epsilon, \quad \ldots \quad (265)$$

and this, by Eqs. (256), becomes

$$N_i x + M_i y + L_i z = \epsilon. \quad \ldots \quad (266)$$

Dividing both members by Rk, we have

$$\frac{N_i}{Rk}x + \frac{M_i}{Rk}y + \frac{L_i}{Rk}z = \frac{\epsilon}{Rk} = p. \quad \ldots \quad (267)$$

The sum of the squares of the coefficients of the variables being unity, these coefficients are the cosines of the angles which the normal to the plane makes with the co-ordinate axes, and $\frac{\epsilon}{Rk}$ is the perpendicular distance from the centre of mass to the plane. We therefore see that the tangent plane at the instantaneous pole is parallel to the invariable plane, and that these two planes are separated by the constant perpendicular distance, p.

Hence the ellipsoid *rolls, without sliding*, on a tangent plane parallel to the plane of resultant rotation of the system, and at a fixed distance p from the centre of the ellipsoid. As different points of the ellipsoid come successively into the tangent plane, the semi-diameters which join them with the centre become in turn the instantaneous axis. The locus of the tangent points on the ellipsoid is called the *Polhode*, and in the general case is a curve of double-curvature. The locus of the points of contact on the tangent plane is necessarily a plane curve, and is called the *Herpolhode*. If we imagine all points of the polhode joined with the centre of the ellipsoid by the various semi-diameters, and all points of the herpolhode with the same point by the various instantaneous axes, we will have two cones; the former called the *rolling cone*, described about a principal axis of the ellipsoid, and the latter called the *directing cone*, about the invariable axis; at any instant they are tangent to each other along the instantaneous axis.

109. *The Rolling Cone.*—Dividing both members of Eq. (261) by e^2, we have

$$A^2\frac{\omega_x^2}{e^2} + B^2\frac{\omega_y^2}{e^2} + C^2\frac{\omega_z^2}{e^2} = \frac{1}{p^{\prime 2}} \quad . \quad . \quad . \quad (268)$$

But, Eqs. (264),

$$\frac{\omega_x^2}{e^2} = x^{\prime\prime 2}; \quad \frac{\omega_y^2}{e^2} = y^{\prime\prime 2}; \quad \frac{\omega_z^2}{e^2} = z^{\prime\prime 2}, \quad \ldots \quad (269)$$

and Eq. (268) becomes

$$A^2 x^{\prime\prime 2} + B^2 y^{\prime\prime 2} + C^2 z^{\prime\prime 2} = \frac{1}{p^{\prime 2}}, \quad \ldots \quad (270)$$

which is an equation of condition for points of the polhode.

Since the instantaneous pole is on the ellipsoid, we have also the condition, Eq. (212),

$$A x^{\prime\prime 2} + B y^{\prime\prime 2} + C z^{\prime\prime 2} = 1, \quad \ldots \quad (271)$$

and dividing by p^2 we get

$$\frac{Ax'^2}{p^2} + \frac{By'^2}{p^2} + \frac{Cz'^2}{p^2} = \frac{1}{p^2}. \quad \ldots \quad (272)$$

Subtracting Eq. (270) from Eq. (272) and omitting accents, we have

$$A\left(\frac{1}{p^2} - A\right)x^2 + B\left(\frac{1}{p^2} - B\right)y^2 + C\left(\frac{1}{p^2} - C\right)z^2 = 0, \quad (273)$$

which expresses the relations existing between the co-ordinates of any point of the polhode. But we see that this equation is satisfied by the co-ordinates of the origin, and also by any set of values which bear a constant ratio to the corresponding co-ordinates of any point of the polhode; that is, the equation is satisfied by the co-ordinates of all points of the instantaneous axis in all of its positions. It is therefore the equation of the rolling cone.

By replacing A, B, C within the parentheses by $\frac{1}{a^2}$, $\frac{1}{b^2}$, $\frac{1}{c^2}$, respectively, $a > b > c$ being the semi-axes of the ellipsoid, the equation of the rolling cone becomes

$$A\left(\frac{1}{p^2} - \frac{1}{a^2}\right)x^2 + B\left(\frac{1}{p^2} - \frac{1}{b^2}\right)y^2 + C\left(\frac{1}{p^2} - \frac{1}{c^2}\right)z^2 = 0. \quad (274)$$

The position and character of this cone depend on the values of the constant p.

110. *Discussion of the Rolling Cone.*

(1) Let $p = a$. Eq. (274) becomes

$$B\left(\frac{1}{b^2} - \frac{1}{a^2}\right)y^2 + C\left(\frac{1}{c^2} - \frac{1}{a^2}\right)z^2 = 0, \quad \ldots \quad (275)$$

which is satisfied only by

$$x = \frac{0}{0}, \quad y = 0 \quad \text{and} \quad z = 0.$$

Hence the axis of x is the rolling cone, the directing cone, the invariable axis, the instantaneous axis, and the longest principal axis of the body. The body rotates uniformly about this axis.

(2) Let $a > p > b$. The second and third terms of Eq. (274) then have the same sign; x is then the axis of the cone, and the sections of the cone normal to x are ellipses.

(3) Let $p = b$. Eq. (274) becomes

$$A\left(\frac{1}{b^2} - \frac{1}{a^2}\right)x^2 - C\left(\frac{1}{c^2} - \frac{1}{b^2}\right)z^2 = 0, \quad \ldots \quad (276)$$

and we have

$$x = \pm \frac{a^2}{c^2}\sqrt{\frac{b^2 - c^2}{a^2 - b^2}}z; \quad \ldots \ldots \quad (277)$$
$$y = \ ;$$

which are the equations of two planes equally inclined to the principal plane ab of the ellipsoid and intersecting in the mean principal axis. They cut from the ellipsoid two equal ellipses, which are called the *critical ellipses*, or *separating polhodes*, of the central ellipsoid. The semi-axes of the critical ellipses are

$$\sqrt{a^2 + c^2 - \frac{a^2 c^2}{b^2}} \quad \text{and} \quad b. \ldots \ldots \quad (278)$$

(4) Let $b > p > c$. Then the first and second terms of Eq. (274) have the same sign; z is then the axis of the cone, and the sections of the cone normal to z are ellipses.

(5) Let $p = c$. Then

$$x = 0, \quad y = 0, \quad z = \frac{0}{0},$$

and the axis of z is the rolling cone, the directing cone, the invariable axis, the instantaneous axis, and the shortest principal axis of the body. The body rotates uniformly about this axis.

III. *The Polhode and Herpolhode.*—In the 1st and 5th cases the polhode and herpolhode are points.

In the 2d and 4th cases the polhode is in general a curve of

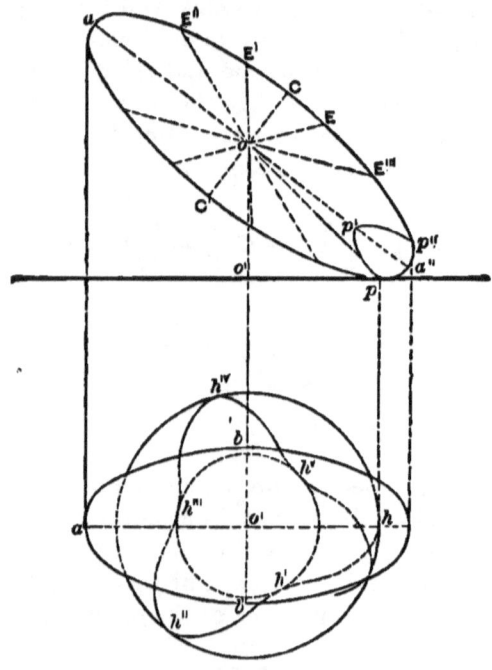

FIG. 44.

double curvature, and the herpolhode is a wavy curve, as indicated in Fig. 44.

In Fig. 44 let $a > p > b$, take the tangent plane to be the horizontal plane, and assume the vertical plane parallel to the plane of the longest and shortest principal axes. Then the longest, shortest and mean axes are projected equal to themselves in aa'', CC' and bb', respectively; $pp'p''$ is the vertical projection of the polhode, and $hh'h''h'''$, etc., is the herpolhode; the invariable axis is projected in OO'' and O', and the instantaneous axis in Op and $O'h$; OE' and OE are traces of the cyclic planes, and OE'' and OE''' are traces of the planes of the critical ellipses.

The maxima and minima values of δ recur when the vertices of the polhode come into the tangent plane. These vertices are the intersections of the polhode by the principal planes of the ellipsoid. The maxima and minima values of the radius vector of the herpolhode correspond to those of δ, and hence this curve will lie between the circumferences of two circles whose common centre is O', and whose radii are $O'h$ and $O'h'$, corresponding to the greatest and least values of δ. If the angle included between two consecutive maximum radii-vectores of the herpolhode be commensurable with a right angle, the curve will be retraced after a certain number of complete turns, and the herpolhode will be a closed curve. If this angle be incommensurable with a right angle, the instantaneous pole will never retrace its former path on the tangent plane.

The general value of the radius vector of the herpolhode is given by the equation

$$\rho^2 = \delta^2 - \frac{\epsilon^2}{K^2 k^2} = \delta^2 - p^2; \quad \ldots \ldots \quad (279)$$

from which, together with $ds = ds'$, in which ds is the length of its elementary arc, while ds' is that of the corresponding element of the polhode, the curve may be found.

In the 3d case the herpolhode becomes a spiral whose pole is O' and whose maximum radius vector is

$$\left(\frac{a^2 b^2 + b^2 c^2 - a^2 c^2}{b^2} - b^2\right)^{\frac{1}{2}}, \quad \ldots \ldots \quad (280)$$

corresponding to an instantaneous axis coincident with the semi-transverse axis of the critical ellipse.

A complete analysis shows that if the rotation be about b it will remain so, but if the body be started to rotate about any other diameter of the critical ellipse it will require an infinite time for the instantaneous axis to reach b.

112. *Permanent Axes*.—To explain what is understood by *permanent* axes of rotation, let the initial impulsion be such as to cause the body to rotate about any of the *central principal axes*. Then, as we have seen, the polhode and herpolhode are coincident points, and the instantaneous, principal and invariable axes are initially coincident, and will continue so during the entire motion. For this reason the *three central principal axes* are called *permanent* axes of rotation; the discussion shows that they are the only lines which have this property of permanence.

113. *Stability of Rotation*.—The critical ellipses divide the surface of the ellipsoid into four areas, two surrounding the extremities of the shortest axis, and the other two the extremities of the longest axis. Within these areas the corresponding polhodes appertaining to each axis are found.

If the impulsion be such as to develop initially an instantaneous axis very near the shortest axis of the ellipsoid, the corresponding polhode will be a small curve surrounding its extremity, and the successive positions of the instantaneous axis meeting the surface in this polhode will never depart very far from the shortest axis. It will periodically return to its initial position in the body, after passing through its two maximum and minimum displacements with respect to the shortest axis. This will likewise be true for any initial or subsequent additional impulsion which causes the polhode to lie within the assumed area. If the initial impulsion develop an instantaneous axis whose polhode surrounds the longest axis, the successive instantaneous axes will be related to that axis in a precisely similar manner. But since the mean principal axis lies in the planes of the critical ellipses, any instantaneous axis not in one of these planes, however near it may be to the mean axis, will belong to a polhode surrounding the longest or shortest axis, and will depart far from the mean axis. The longest and shortest axes are therefore called *stable axes* of rotation, and the *mean axis* an *unstable axis*.

When $a = b$ the central ellipsoid is an oblate spheroid, and the critical ellipses unite in the equator. The areas reduce to two, surrounding the axis of the spheroid, which is the only stable axis of rotation, and the polhodes are circles whose pole is the extremity of the axis.

When $b = c$ the central ellipsoid becomes a prolate spheroid, the critical ellipses unite in its equator, and the axis of the spheroid is the only axis of stability. An elongated rifled projectile is such a body, and its axis is the only stable axis of rotation. A very great initial angular velocity is usually given to it about this axis, so that the influences which modify its angular velocity, either as to amount or change of axis, while the projectile is describing its trajectory, will be comparatively so minute as not to cause the instantaneous axis to depart sensibly from this axis of stability.

Rotation Due to Incessant Forces.

114. *Euler's Equations of Rotation.*—Differentiating Eqs. (251), we have

$$\left. \begin{aligned} \frac{d^2 x'}{dt^2} &= z'\frac{d\omega_y}{dt} - y'\frac{d\omega_z}{dt} + \omega_y(y'\omega_x - x'\omega_y) - \omega_z(x'\omega_z - z'\omega_x); \\ \frac{d^2 y'}{dt^2} &= x'\frac{d\omega_z}{dt} - z'\frac{d\omega_x}{dt} + \omega_z(z'\omega_y - y'\omega_z) - \omega_x(y'\omega_x - x'\omega_y); \\ \frac{d^2 z'}{dt^2} &= y'\frac{d\omega_x}{dt} - x'\frac{d\omega_y}{dt} + \omega_x(x'\omega_z - z'\omega_x) - \omega_y(z'\omega_y - y'\omega_z); \end{aligned} \right\} . (281)$$

and, substituting in Eqs. (R_m), the latter become, after omitting accents,

$$\begin{aligned}
\Sigma m\frac{xd^2y - yd^2x}{dt^2} &= \Sigma m(x^2+y^2)\frac{d\omega_z}{dt} + \Sigma m(x^2-y^2)\omega_x\omega_y \\
&\quad - \Sigma mxy(\omega_x^2-\omega_y^2) - \Sigma myz\left(\frac{d\omega_y}{dt}+\omega_x\omega_z\right) \\
&\quad - \Sigma mxz\left(\frac{d\omega_x}{dt}-\omega_y\omega_z\right) = L; \\
\Sigma m\frac{zd^2x - xd^2z}{dt^2} &= \Sigma m(x^2+z^2)\frac{d\omega_y}{dt} + \Sigma m(z^2-x^2)\omega_z\omega_x \\
&\quad - \Sigma mxz(\omega_z^2-\omega_x^2) - \Sigma mxy\left(\frac{d\omega_x}{dt}+\omega_y\omega_z\right) \\
&\quad - \Sigma myz\left(\frac{d\omega_z}{dt}-\omega_x\omega_y\right) = M; \\
\Sigma m\frac{yd^2z - zd^2y}{dt^2} &= \Sigma m(y^2+z^2)\frac{d\omega_x}{dt} + \Sigma m(y^2-z^2)\omega_y\omega_z \\
&\quad - \Sigma myz(\omega_y^2-\omega_z^2) - \Sigma mxz\left(\frac{d\omega_z}{dt}+\omega_x\omega_y\right) \\
&\quad - \Sigma mxy\left(\frac{d\omega_y}{dt}-\omega_x\omega_z\right) = N.
\end{aligned} \quad (282)$$

Equations R_m and (281) are true whatever be the directions of the axes x', y', z'. Let the co-ordinate axes be the principal axes of the body; then Eqs. (282) may be written

$$\begin{aligned}
C\frac{d\omega_z}{dt} - (A-B)\omega_x\omega_y &= L; \\
B\frac{d\omega_y}{dt} - (C-A)\omega_x\omega_z &= M; \\
A\frac{d\omega_x}{dt} - (B-C)\omega_y\omega_z &= N.
\end{aligned} \quad \ldots \quad (283)$$

These equations are known as *Euler's equations of rotation,* and the values of ω_x, ω_y and ω_z may be found from them when integration is possible. But since the principal axes conform to the motion of the body, the complete solution requires that the position of these axes at any instant shall be determined with respect to the axes fixed in direction.

These equations, like Eqs. (R_m), also apply to rotation about a fixed axis or a fixed point.

115. *Auxiliary Angles.*—Let X, Y, Z, Fig. 45, be the axes fixed in direction, and X', Y', Z' the principal axes of the body. Conceive a sphere to be described about O with unit radius. Its intersections with the co-ordinate planes are xy, xz, yz, $x'y'$, $x'z'$, $y'z'$, and the intersection of the planes XY and $X'Y'$ is ON. Assume the notation

$$\begin{rcases} X'NY = ZOZ' = \theta; \\ XON = \psi; \\ X'ON = \phi. \end{rcases} \quad (284)$$

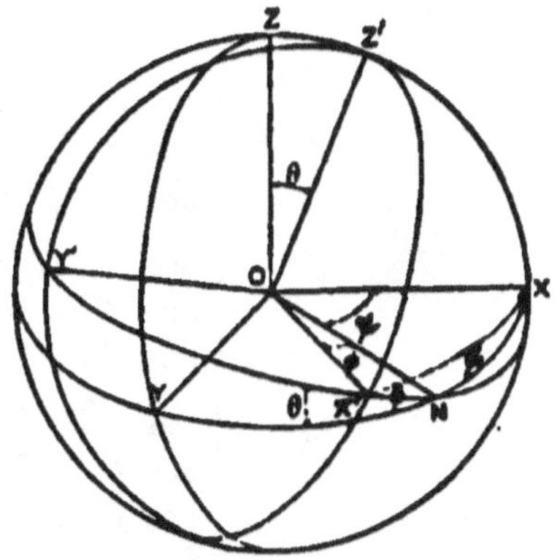

ON is called the *line of the nodes*, θ the *obliquity*, and ϕ the *precession*.

FIG. 45.

Taking Z' positive when $Z'OZ < 90°$, we have from the spherical triangles of the figure, considering N as a vertex in each,

$$\begin{rcases} \cos xOx' = \cos \phi \cos \psi - \sin \phi \sin \psi \cos \theta; \\ \cos xOy' = -\sin \phi \cos \psi - \cos \phi \sin \psi \cos \theta; \\ \cos xOz' = \sin \psi \sin \theta; \\ \cos yOx' = \cos \phi \sin \psi + \sin \phi \cos \psi \cos \theta, \\ \cos yOy' = -\sin \phi \sin \psi + \cos \phi \cos \psi \cos \theta; \\ \cos yOz' = -\cos \psi \sin \theta; \\ \cos zOx' = \sin \phi \sin \theta; \\ \cos zOy' = \cos \phi \sin \theta; \\ \cos zOz' = \cos \theta. \end{rcases} \quad (285)$$

116. The component angular velocities about ON, Z and Z' are frequently used as auxiliary angular velocities for the determination of ω_x, ω_y and ω_z; they are respectively $\frac{d\theta}{dt}$, $\frac{d\psi}{dt}$ and $\frac{d\phi}{dt}$. The first is called the *nutation of obliquity*, and the second the *precessional velocity*. The latter is *direct* when ψ increases with the time, and *retrograde* when it decreases.

For the angles which the axes Z, Z' and the line ON make with the axes X', Y', Z' we have

$$\left.\begin{array}{l} \left\{\begin{array}{l} \cos ZOX' = \sin\phi \sin\theta; \\ \cos ZOY' = \cos\phi \sin\theta; \\ \cos ZOZ' = \cos\theta; \end{array}\right. \\ \left\{\begin{array}{l} \cos Z'OX' = 0; \\ \cos Z'OY' = 0; \\ \cos Z'OZ' = 1; \end{array}\right. \\ \left\{\begin{array}{l} \cos NOX' = \cos\phi; \\ \cos NOY' = -\sin\phi; \\ \cos NOZ' = 0; \end{array}\right. \end{array}\right\} \quad \ldots \quad (286)$$

and hence, by the principle of the parallelopipedon of angular velocities, we have

$$\left.\begin{array}{l} \omega_x = \dfrac{d\theta}{dt}\cos\phi + \dfrac{d\psi}{dt}\sin\phi\sin\theta; \\ \omega_y = -\dfrac{d\theta}{dt}\sin\phi + \dfrac{d\psi}{dt}\cos\phi\sin\theta; \\ \omega_z = \dfrac{d\phi}{dt} + \dfrac{d\psi}{dt}\cos\theta. \end{array}\right\} \quad \ldots \quad (287)$$

117. *The Gyroscope*.—The problem of the gyroscope illustrates this subject. It may be stated thus:

Find the circumstances of motion of a solid of revolution about a fixed point on its axis, it having been given an initial

ROTATION DUE TO INCESSANT FORCES.

rotation about its axis and then left to the action of its own weight.

Let O, Fig. 46, be the fixed point, Ox, Oy, Oz the fixed axes, and Ox', Oy', Oz' the principal axes at O, Oz being taken vertical and positive upward. Oz' is the axis of revolution of the body. Let h be the distance from O to the centre of gravity, and assume $A = B$ and $B < C$.

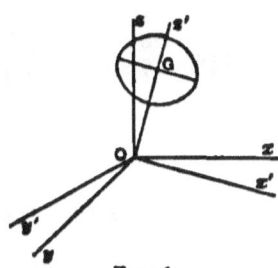

Fig. 46.

The components of the weight in the directions of Ox', Oy' and Oz' are

$$\left. \begin{array}{l} X' = mg \cos(180° - zOx') = -mg \cos zOx'; \\ Y' = mg \cos(180° - zOy') = -mg \cos zOy'; \\ Z' = mg \cos(180° - zOz') = -mg \cos zOz'; \end{array} \right\} \quad (288)$$

and the component moments of the weight with respect to the same axes are, Eqs. (286),

$$\left. \begin{array}{l} N = Z'y' - Y'z' = mgh \cos zOy' = mgh \sin \theta \cos \phi; \\ M = X'z' - Z'x' = -mgh \cos zOx' = -mgh \sin \theta \sin \phi; \\ L = Y'x' - X'y' = 0. \end{array} \right\} \quad (289)$$

Substituting these values in Euler's Eqs. (283), we have

$$\left. \begin{array}{l} A\dfrac{d\omega_x}{dt} - (A-C)\omega_y\omega_z = N = mgh \sin \theta \cos \phi; \\ A\dfrac{d\omega_y}{dt} + (A-C)\omega_x\omega_z = M = -mgh \sin \theta \sin \phi; \\ C\dfrac{d\omega_z}{dt} = L = 0. \end{array} \right\} \quad (290)$$

Integrating the last equation, we have

$$\omega_z = \text{a constant} = n. \quad \ldots \ldots \quad (291)$$

Multiplying the first of Eqs. (290) by ω_x and the second by ω_y, we have by addition, Eqs. (287),

$$A\left(\omega_x \frac{d\omega_x}{dt} + \omega_y \frac{d\omega_y}{dt}\right) = mgh \sin\theta(\omega_x \cos\phi - \omega_y \sin\phi) \\ = mgh \sin\theta \frac{d\theta}{dt}. \quad \Big\} \quad (292)$$

Integrating this equation, we have

$$\tfrac{1}{2}A(\omega_x^2 + \omega_y^2) = mgh(\cos\theta_0 - \cos\theta), \quad \cdot \quad \cdot \quad (293)$$

in which θ_0 is the initial angle zOz'.

Adding the initial kinetic energy of rotation $\tfrac{1}{2}Cn^2$, we have

$$\tfrac{1}{2}A(\omega_x^2 + \omega_y^2) + \tfrac{1}{2}Cn^2 = mgh(\cos\theta_0 - \cos\theta) + \tfrac{1}{2}Cn^2. \quad (294)$$

Since $mgh(\cos\theta_0 - \cos\theta)$ is the work of the weight while G falls over the distance $h(\cos\theta_0 - \cos\theta)$, we see that Eq. (294) expresses the theorem of kinetic energy of rotation.

118. From Eqs. (287) we readily get

$$\omega_x^2 + \omega_y^2 = \frac{d\theta^2}{dt^2} + \sin^2\theta \frac{d\psi^2}{dt^2}; \quad \cdot \quad \cdot \quad \cdot \quad \cdot \quad (295)$$

which substituted in Eq. (294) gives

$$A\frac{d\theta^2}{dt^2} + A\sin^2\theta \frac{d\psi^2}{dt^2} = 2mgh(\cos\theta_0 - \cos\theta). \quad \cdot \quad (296)$$

Multiplying the first and second of Eqs. (290) by $\sin\phi$ and $\cos\phi$ respectively, we have, by addition,

$$A\left(\sin\phi\frac{d\omega_x}{dt}+\cos\phi\frac{d\omega_y}{dt}\right)+n(C-A)[\omega_y\sin\phi-\omega_x\cos\phi]=0, \quad (297)$$

and, since Eqs. (287) give

$$\omega_y\sin\phi-\omega_x\cos\phi=-\frac{d\theta}{dt}, \quad\ldots\quad (298)$$

Eq. (297) reduces to

$$A\left(\sin\phi\frac{d\omega_x}{dt}+\cos\phi\frac{d\omega_y}{dt}\right)+nA\frac{d\theta}{dt}-nC\frac{d\theta}{dt}=0. \quad (299)$$

We have also, from Eqs. (287),

$$\omega_x\sin\phi+\omega_y\cos\phi=\sin\theta\frac{d\psi}{dt}. \quad\ldots\quad (300)$$

Differentiating this last, dividing by dt and reducing by the relations of Eqs. (287), we have

$$\sin\phi\frac{d\omega_x}{dt}+\cos\phi\frac{d\omega_y}{dt}=\cos\theta\frac{d\theta}{dt}\frac{d\psi}{dt}+\sin\theta\frac{d^2\psi}{dt^2}$$

$$-(\omega_x\cos\phi-\omega_y\sin\phi)\frac{d\phi}{dt}$$

$$=\cos\theta\frac{d\theta}{dt}\frac{d\psi}{dt}+\sin\theta\frac{d^2\psi}{dt^2}-\left(n-\cos\theta\frac{d\psi}{dt}\right)\frac{d\theta}{dt}$$

$$=2\cos\theta\frac{d\theta}{dt}\frac{d\psi}{dt}+\sin\theta\frac{d^2\psi}{dt^2}-n\frac{d\theta}{dt}. \quad (301)$$

Substituting this value in Eq. (299), we have

$$A\left(2\cos\theta\frac{d\theta}{dt}\frac{d\psi}{dt}+\sin\theta\frac{d^2\psi}{dt^2}\right)-Cn\frac{d\theta}{dt}=0; \quad\ldots\quad (302)$$

which, after multiplying by $\sin \theta$ and integrating, gives

$$A \sin^2 \theta \frac{d\psi}{dt} + Cn \cos \theta = \text{a constant} = h'; \quad . \quad . \quad (303)$$

or between limits,

$$A \sin^2 \theta \frac{d\psi}{dt} = Cn(\cos \theta_0 - \cos \theta). \quad . \quad . \quad . \quad . \quad (304)$$

From the last of Eqs. (287) we also have, since $\omega_z = n$,

$$\frac{d\phi}{dt} + \frac{d\psi}{dt} \cos \theta = n. \quad . \quad . \quad . \quad . \quad . \quad . \quad (305)$$

Eqs. (296), (304) and (305), viz.,

$$\left. \begin{array}{l} A\dfrac{d\theta^2}{dt^2} + A \sin^2 \theta \dfrac{d\psi^2}{dt^2} = 2mgh (\cos \theta_0 - \cos \theta), \\[4pt] A \sin^2 \theta \dfrac{d\psi}{dt} = Cn (\cos \theta_0 - \cos \theta), \\[4pt] \dfrac{d\phi}{dt} + \dfrac{d\psi}{dt} \cos \theta = n, \end{array} \right\} \quad (306)$$

are the differential equations of motion of the gyroscope. From them the values of θ, ψ and ϕ, which give the position of the body at any instant, may be found in terms of known quantities.

119. Square the second of Eqs. (306) and multiply the first by $A \sin^2 \theta$ and eliminate $\dfrac{d\psi^2}{dt^2}$ from the resulting equations by subtraction; then solve with respect to $\dfrac{d\theta}{dt}$, and we have, for the nutational velocity,

$$\frac{d\theta}{dt} = \pm \frac{(\cos\theta_0 - \cos\theta)^{\frac{1}{2}}[2mghA\sin^2\theta - C^2n^2(\cos\theta_0 - \cos\theta)]^{\frac{1}{2}}}{A\sin\theta}. \quad (307)$$

Any value of θ less than θ_0 makes $\frac{d\theta}{dt}$ imaginary, and therefore the centre of gravity can never get above its initial position. And since $\frac{d\theta}{dt}$ is imaginary when $\theta = 180°$, the centre of gravity can never reach the vertical through the fixed point.

Making $\frac{d\theta}{dt} = 0$ in Eq. (307), we find that the resulting equation may be satisfied by $\theta = \theta_0$, and also by the two roots of the equation

$$2mghA\sin^2\theta - C^2n^2(\cos\theta_0 - \cos\theta) = 0. \quad (308)$$

One of these roots gives a maximum value of θ, which we will call θ_1. The other root gives $\cos\theta > 1$, and hence corresponds to no angle. Thus we see that the body falls from its initial position until $\theta = \theta_1$, then rises until $\theta = \theta_0$, and continues to oscillate between these two values. The integral of dt (Eq. 307) between the limits θ_0 and θ_1 evidently gives half the time of a complete nutational oscillation.

The precessional velocity given by the second of Eqs. (306) is

$$\frac{d\psi}{dt} = \frac{Cn(\cos\theta_0 - \cos\theta)}{A\sin^2\theta}; \quad \ldots \quad (309)$$

and it is zero when $\theta = \theta_0$, and a maximum when $\theta = \theta_1$. It is direct or retrograde according as n is positive or negative.

Combining the motions in nutation and precession, we find that the horizontal projection of the path of the centre of gravity lies between two concentric circumferences whose radii are $h \sin \theta_0$ and $h \sin \theta_1$, and is tangent to the outer and normal to the inner circumference.

From Eq. (308) we have

$$\sin \theta = Cn \sqrt{\frac{\cos \theta_0 - \cos \theta_1}{2mghA}}; \quad \ldots \quad (310)$$

from which we see that $\theta_1 - \theta_0$ may, by increasing the value of n, be made less than any assignable quantity. In the common gyroscope we can give n such a value that the eye can detect neither the vertical motion nor the variation in the precessional velocity.

This discussion gives only the general character of the motion. The complete solution of the problem requires the integration of Eqs. (306), and this involves methods not given in the course of mathematics at the Military Academy.

Impact.

120. When two bodies collide there is a transfer of energy from one to the other, by which changes in the velocities of both bodies, and in their form and volume, are effected. The impact, though ordinarily said to be instantaneous, requires a finite time for its completion. When after collision we examine the surfaces of two ivory balls which have been previously oiled, we notice that their areas of contact during collision must have been very much greater than when they simply rest against each other. Hence the distance between their centres of mass during impact must be less than the sum of the radii of the spheres, and the intensity of the mutual pressure of the colliding bodies evidently varies by continuity from zero to a maximum and then to zero again. The instant of nearest approach of their centres separates the period of *compression* from that of *restitution*, and at that instant their centres have the same velocity.

Actual solids possess a certain degree of elasticity of form and volume, by which they regain approximately their original form and volume when the impact has ended. If C be the intensity of the impulsion producing compression, R that which restores the form and volume, and e their ratio, we have

$$R = eC; \quad \ldots \ldots \ldots \quad (311)$$

e is called the *coefficient of restitution*, and in actual bodies is less than unity and greater than zero.

121. *Direct and Central Impact.*—Let a spherical mass m, moving with a velocity v, collide with a similar mass m', whose velocity is u in the same direction, and let w be their common velocity at the instant of nearest approach of their centres; then, taking velocities in opposite directions to have opposite signs, we have

$$C = m(v - w) = m'(w - u), \quad \ldots \ldots \quad (312)$$

whence

$$C = \frac{mm'}{m + m'}(v - u) \quad \ldots \ldots \ldots \quad (313)$$

and

$$R = Ce = \frac{mm'}{m + m'}(v - u)e. \quad \ldots \ldots \quad (314)$$

Let V and U be the final velocities of m and m', and we have

$$R + C = C(1 + e) = \frac{mm'}{m + m'}(v - u)(1 + e)$$
$$= m(v - V) = m'(U - u). \quad \ldots \quad (315)$$

Hence

$$\left.\begin{aligned}V &= v - \frac{m'}{m+m'}(v-u)(1+e); \\ U &= u + \frac{m}{m+m'}(v-u)(1+e).\end{aligned}\right\} \quad \ldots \quad (316)$$

The limits of these values, found by making $e = 1$ and $e = 0$, are

$$\left.\begin{aligned}V_1 &= v - \frac{2m'}{m+m'}(v-u), \\ U_1 &= u + \frac{2m}{m+m'}(v-u),\end{aligned}\right\} \quad \ldots \ldots \quad (317)$$

and

$$\left.\begin{aligned}V_2 &= v - \frac{m'}{m+m'}(v-u) = \frac{mv+m'u}{m+m'}, \\ U_2 &= u + \frac{m}{m+m'}(v-u) = \frac{mv+m'u}{m+m'}.\end{aligned}\right\} \quad (318)$$

In actual cases e is a constant to be determined by experiment. Whatever its value may be, it is readily seen from Eqs. (316) that the sum of the momenta of the two bodies after collision is equal to that before collision; therefore no momentum is destroyed by the impact.

The sum of the kinetic energies of the masses before and after impact are respectively

$$\tfrac{1}{2}(mv^2 + m'u^2) \quad \ldots \ldots \ldots \quad (319)$$

and

$$\tfrac{1}{2}(mv^2 + m'u^2) - \tfrac{1}{2}\frac{mm'}{m+m'}(v-u)^2(1-e^2). \quad \ldots \quad (320)$$

Whence we see that a loss of kinetic energy always accompanies the impact of actual masses.

122. *Oblique Impact.*—If the paths of the colliding bodies be oblique to each other, their velocities may be resolved into components in the directions of the common tangent and normal at the point of impact. C and R will depend only on the normal components, the tangential components having no effect on the impact. The final velocity of each body will therefore be given by the resultant of the changed normal component and the unchanged tangential component.

123. If m' be very great with respect to m and at rest we have the case of impact against a fixed obstacle. Then if ϕ be the angle of incidence, Fig. 47, or that which the direction of the path of m makes with the normal to the deviating surface at the point of m' impact, we have for the component momenta before compression

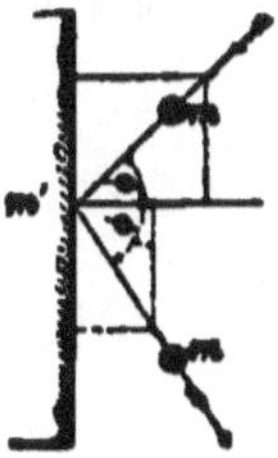

$$mv \sin \phi \quad \text{and} \quad mv \cos \phi,$$

Fig. 47.

and for those after restitution

$$mv \sin \phi \quad \text{and} \quad - mev \cos \phi;$$

the resultant of the latter being

$$mv \sqrt{\sin^2\phi + e^2 \cos^2\phi}.$$

If the angle of reflection is ϕ' we have

$$\tan \phi' = - \frac{\sin \phi}{e \cos \phi} = - \frac{\tan \phi}{e}, \quad \ldots \quad (321)$$

which varies between $- \tan \phi$ for $e = 1$, and ∞ for $e = 0$. Hence in all actual cases of impact the path of the reflected body will make an angle with the normal greater than that of incidence and less than 90°, depending on the value of e for the bodies considered.

Axis of Spontaneous Rotation.

124. A *spontaneous* axis is a right line *fixed in space*, about which a free body rotates during impact while the centre of mass of the body is in motion. Its position and the necessary conditions for its development are derived from Eqs. (117). Dividing these by dt, and replacing the component angular velocities about the centre of mass by their symbols, we have

$$\left. \begin{array}{l} \dfrac{dx}{dt} = \dfrac{dx_0}{dt} + z'\omega_y - y'\omega_z; \\[4pt] \dfrac{dy}{dt} = \dfrac{dy_0}{dt} + x'\omega_z - z'\omega_x; \\[4pt] \dfrac{dz}{dt} = \dfrac{dz_0}{dt} + y'\omega_x - x'\omega_y; \end{array} \right\} \quad \dots \quad (322)$$

in which the first members are the component velocities of any molecule of the body with reference to a set of axes fixed in space, and the first terms of the second member are the component velocities of the centre of mass with respect to the fixed origin, when the centre of mass is taken as the movable origin.

From Eqs. (T_m') we have

$$\left. \begin{array}{l} \dfrac{dx_0}{dt} = V_x = \dfrac{X_t}{M}; \\[4pt] \dfrac{dy_0}{dt} = V_y = \dfrac{Y_t}{M}; \\[4pt] \dfrac{dz_0}{dt} = V_z = \dfrac{Z_t}{M}; \end{array} \right\} \quad \dots \quad (323)$$

in which X_t, Y_t, Z_t are the component intensities of the resultant impulsion in the direction of the co-ordinate axes.

Substituting these values in Eqs. (322), we have

$$\left.\begin{aligned}\frac{dx}{dt} &= \frac{X_i}{M} + z'\omega_y - y'\omega_z; \\ \frac{dy}{dt} &= \frac{Y_i}{M} + x'\omega_z - z'\omega_x; \\ \frac{dz}{dt} &= \frac{Z_i}{M} + y'\omega_x - x'\omega_y.\end{aligned}\right\} \quad \ldots \ldots (324)$$

For all points at rest with respect to the fixed origin we have the conditions

$$\frac{dx}{dt} = \frac{dy}{dt} = \frac{dz}{dt} = 0, \ \ldots \ldots (325)$$

and Eqs. (324) become

$$\left.\begin{aligned}\frac{X_i}{M} + z'\omega_y - y'\omega_z &= 0; \\ \frac{Y_i}{M} + x'\omega_z - z'\omega_x &= 0; \\ \frac{Z_i}{M} + y'\omega_x - x'\omega_y &= 0;\end{aligned}\right\} \quad \ldots \ldots (326)$$

which will be the equations of a right line fixed in space when

$$X_i\omega_x + Y_i\omega_y + Z_i\omega_z = 0. \ \ldots \ldots (327)$$

Dividing Eq. (327) by $R_i\omega$, we have

$$\frac{X_i}{R_i}\frac{\omega_x}{\omega} + \frac{Y_i}{R_i}\frac{\omega_y}{\omega} + \frac{Z_i}{R_i}\frac{\omega_z}{\omega} = 0, \ \ldots \ldots (328)$$

which expresses the condition that the action-line of the resultant impulsion is perpendicular to the instantaneous axis; hence

140   MECHANICS OF SOLIDS.

we conclude that *a spontaneous axis will be developed when a body is so struck as to make the instantaneous axis perpendicular to the resultant impulsion;* otherwise Eqs. (326) are the equations of a single point which alone is at rest for the instant. We see also by comparing Eqs. (326) with Eqs. (252) that *the spontaneous axis is always parallel to the instantaneous axis.*

125. It is readily seen from Eq. (327) that the required condition can be satisfied only when at least one factor of each term is zero; that is, either *when the line of impact lies in a central principal plane or when it is parallel to a central principal axis.* The discussion is the same for all cases.

126. Let the line of resultant impact, Fig. 48, be parallel to the principal axis y' and lie in the principal plane $x'y'$, and let $R_{\prime} = MV$ be the intensity of the resultant impulsion; then we have

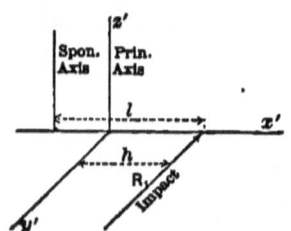

Fig. 48.

$$X_{\prime} = 0; \quad Y_{\prime} = MV; \quad Z_{\prime} = 0;$$
$$\omega_x = 0; \quad \omega_y = 0; \quad \omega_z = \frac{MVh}{C}; \quad (329)$$

h being the lever arm of the impulsion with respect to the axis z'; Eq. (327) is satisfied, and the equations of the spontaneous axis developed by the impact are

$$y' = 0; \quad z' = \frac{0}{0}; \quad x' = -\frac{C}{Mh} = -\frac{Mk_{\prime}^2}{Mh} = -\frac{k_{\prime}^2}{h}; \quad (330)$$

which are those of a right line parallel to the principal axis z', intersecting the axis x' at a distance from the centre of mass equal to $-\frac{k_{\prime}^2}{h}$.

Let l be the perpendicular distance between the line of impact and the spontaneous axis; then we have

$$l = h + \frac{k_{\prime}^2}{h}; \quad \ldots \ldots \quad (331)$$

whence

$$(l - h)h = k_i'^2. \quad \ldots \ldots (332)$$

Since $k_i'^2$, the square of the principal radius of gyration of the body with respect to the axis s', is constant, we see that the two distances, viz., h from the centre of mass to the line of impact, and $l - h$ from the centre of mass to the spontaneous axis, are reciprocally proportional; hence, as the line of impact recedes from the centre of mass, the spontaneous axis approaches that point, and conversely. When the line of impact passes through the centre of mass the spontaneous axis is at an infinite distance, as it should be, since in this case the body will have motion of translation only.

127. When a spontaneous axis is developed the action-line of the corresponding impulsion is called an *axis of percussion*, and each of its points a *centre of percussion;* the latter term, however, being generally applied to the point in which the axis of percussion intersects the line h. The corresponding point of the spontaneous axis is called a *centre of spontaneous rotation*. The axis of percussion and the spontaneous axis are conjugate lines, each of which implies the other; the positions of both are connected and determined by Eq. (332). For example, the spontaneous axis of a straight rod struck at its extremity in a direction perpendicular to its length is, Eq. (222), given by

$$(l - h)h = (l - a)a = \frac{a^2}{3}; \quad \ldots \ldots (333)$$

whence

$$l - a = \tfrac{1}{3}a, \text{ or } l = \tfrac{2}{3}.2a, \quad \ldots \ldots (334)$$

or is at a distance two thirds of the length of the rod from the line of impact. All elements of the rod beyond the spontaneous axis will have a motion in a direction opposite to that of the impact, and those between the axis and line of impact a motion in the same direction as the impact. This explains the cause of the physical shock experienced when, in striking a ball with a bat or in chopping with an axe, the part held by the hand does not conform to the position of the spontaneous axis corresponding to the line of impact.

128. In Eq. (332) substitute $l - h$ for h, and we have

$$[l - (l - h)](l - h) = h(l - h) = k_i^2. \quad . \quad . \quad . \quad (335)$$

Hence, for parallel impacts, the centre of percussion and the centre of spontaneous rotation are reciprocal and convertible; that is, if the centre of spontaneous rotation become a new centre of percussion, the old centre of percussion will become the new centre of spontaneous rotation.

Constrained Motion.

129. When a rigid surface or curve deflects a body from the free path which any given system of forces would cause it to take, the motion is said to be *constrained*. Let the motion of its centre of mass determine the translation of the body, and, by the principles in Art. 82, we may omit the present consideration of its motion of rotation about that point. Eq. (119) then becomes

$$\left(X - M\frac{d^2x}{dt^2}\right)dx + \left(Y - M\frac{d^2y}{dt^2}\right)dy + \left(Z - M\frac{d^2z}{dt^2}\right)dz = 0. \quad (336)$$

130. *Equations of Constraint.*—Let

$$L = f(x, y, z) = 0 \quad . \quad . \quad . \quad . \quad . \quad (337)$$

be the equation of the surface upon which the centre of mass is constrained to move. Differentiating this equation, we have

$$\frac{dL}{dx}dx + \frac{dL}{dy}dy + \frac{dL}{dz}dz = 0. \quad . \quad . \quad . \quad (338)$$

CONSTRAINED MOTION. 143

Since the path of the centre of mass lies in the given surface, Eqs. (336) and (338) may be combined by considering only those values of dx, dy and dz which are common to the path and surface. To make the terms of these equations quantities of the same kind, multiply the differential equation of the surface by an intensity I. Add the resulting equation to Eq. (336), and we have

$$\left(X - M\frac{d^2x}{dt^2} + I\frac{dL}{dx}\right)dx + \left(Y - M\frac{d^2y}{dt^2} + I\frac{dL}{dy}\right)dy$$
$$+ \left(Z - M\frac{d^2z}{dt^2} + I\frac{dL}{dz}\right)dz = 0. \quad (339)$$

Now, if

$$I\frac{dL}{dx}, \quad I\frac{dL}{dy} \quad \text{and} \quad I\frac{dL}{dz}, \quad \ldots \quad (340)$$

be the rectangular components of a force which, together with the given extraneous forces, will cause the body to remain continually on the geometrical surface whose equation is that of the rigid surface (337), the latter may be supposed removed and the body will be a free body subjected to the action of the component extraneous forces $X + I\frac{dL}{dx}$, etc.; hence, by Eqs. (T$_m$), we will have

$$\left. \begin{array}{l} X + I\frac{dL}{dx} - M\frac{d^2x}{dt^2} = 0; \\ Y + I\frac{dL}{dy} - M\frac{d^2y}{dt^2} = 0; \\ Z + I\frac{dL}{dz} - M\frac{d^2z}{dt^2} = 0. \end{array} \right\} \quad \ldots \quad (341)$$

Eliminating I from these equations, we have

$$\left.\begin{array}{l}\left(X - M\dfrac{d^2x}{dt^2}\right)\dfrac{dL}{dy} - \left(Y - M\dfrac{d^2y}{dt^2}\right)\dfrac{dL}{dx} = 0, \\ \left(X - M\dfrac{d^2x}{dt^2}\right)\dfrac{dL}{dz} - \left(Z - M\dfrac{d^2z}{dt^2}\right)\dfrac{dL}{dx} = 0,\end{array}\right\} \quad (342)$$

which are called the differential *Equations of Constraint*, and from which, with the equation of the surface, the path of the centre of mass and its position at any time may be determined.

131. *The Normal Reaction.*—Let N represent the intensity of the resultant whose component forces are

$$I\frac{dL}{dx}, \quad I\frac{dL}{dy} \quad \text{and} \quad I\frac{dL}{dz},$$

and θ_x, θ_y, θ_z the angles which N makes with the co-ordinate axes; then

$$N = I\sqrt{\frac{dL^2}{dx^2} + \frac{dL^2}{dy^2} + \frac{dL^2}{dz^2}}; \quad \ldots \quad (343)$$

$$\left.\begin{array}{l}\cos\theta_x = \dfrac{I}{N}\dfrac{dL}{dx} = \dfrac{\dfrac{dL}{dx}}{\sqrt{\dfrac{dL^2}{dx^2} + \dfrac{dL^2}{dy^2} + \dfrac{dL^2}{dz^2}}}; \\[2ex] \cos\theta_y = \dfrac{I}{N}\dfrac{dL}{dy} = \dfrac{\dfrac{dL}{dy}}{\sqrt{\dfrac{dL^2}{dx^2} + \dfrac{dL^2}{dy^2} + \dfrac{dL^2}{dz^2}}}; \\[2ex] \cos\theta_z = \dfrac{I}{N}\dfrac{dL}{dz} = \dfrac{\dfrac{dL}{dz}}{\sqrt{\dfrac{dL^2}{dx^2} + \dfrac{dL^2}{dy^2} + \dfrac{dL^2}{dz^2}}}\end{array}\right\} \quad (344)$$

Hence N acts always in the direction of the normal to the deviating surface.

Substituting in Eqs. (341) for $I\frac{dL}{dx}$, etc., their equals $N \cos \theta_x$, etc., we have, after transposing,

$$\left. \begin{array}{l} X - M\dfrac{d^2x}{dt^2} = -N \cos \theta_x; \\ Y - M\dfrac{d^2y}{dt^2} = -N \cos \theta_y; \\ Z - M\dfrac{d^2z}{dt^2} = -N \cos \theta_z. \end{array} \right\} \quad \ldots \quad (345)$$

Squaring and adding, and extracting the square root of the resulting equation, we have

$$\left\{ \left(X - M\frac{d^2x}{dt^2}\right)^2 + \left(Y - M\frac{d^2y}{dt^2}\right)^2 + \left(Z - M\frac{d^2z}{dt^2}\right)^2 \right\}^{\frac{1}{2}} = N. \quad (346)$$

Representing the first member by P and dividing each of Eqs. (345) by (346), we have

$$\left. \begin{array}{l} \dfrac{X - M\dfrac{d^2x}{dt^2}}{P} = -\cos \theta_x; \\ \dfrac{Y - M\dfrac{d^2y}{dt^2}}{P} = -\cos \theta_y; \\ \dfrac{Z - M\dfrac{d^2z}{dt^2}}{P} = -\cos \theta_z. \end{array} \right\} \quad \ldots \quad (347)$$

The first members are the cosines of the angles which the resultant P makes with the co-ordinate axes, and we see therefore that P acts in a direction opposite to the normal of the deviating surface. We also see that P is the resultant of that part of the extraneous forces which generates no momentum, and as its action-line makes an angle of 180° with the normal to the constraining surface, it is the measure of the direct pressure on the surface; the equal intensity N is the equivalent normal reaction of the surface. We therefore see that the force N, which we have apparently introduced, already existed in the reaction of the rigid surface. Hence we conclude that if a body be acted on by any system of extraneous forces and constrained to move upon a rigid surface, the circumstances of motion of translation will be precisely the same as if it were a free body acted upon by a system consisting of the given forces and one whose intensity and direction are those of the normal reaction of the surface. Equations (345) may therefore be employed in problems of constraint, just as Eqs. (T_m) are employed in problems of free motion.

132. Transposing the second terms of the first members of Eqs. (345) to the second members, and multiplying the resulting equations by dx, dy, dz, respectively, we have, by addition,

$$Xdx + Ydy + Zdz + N\left\{\frac{dx}{ds}\cos\theta_x + \frac{dy}{ds}\cos\theta_y + \frac{dz}{ds}\cos\theta_z\right\}ds$$
$$= M\frac{dxd^2x + dyd^2y + dzd^2z}{dt^2}. \quad \ldots \quad (348)$$

which, since

$$\frac{dx}{ds}\cos\theta_x + \frac{dy}{ds}\cos\theta_y + \frac{dz}{ds}\cos\theta_z = 0. \quad \ldots \quad (349)$$

is the cosine of the angle which the normal reaction makes with the tangent to the surface, reduces by integration to

$$\int (Xdx + Ydy + Zdz) = \frac{MV^2}{2} + C, \quad \ldots \quad (350)$$

the equation of energy (126) of a free body under the action of the same forces. Hence the conclusions derived from the theorem of energy in free motion of translation are true in constrained motion also (Arts. 83 and 84); the constraint being supposed to be without friction.

133. To find the value of N, eliminate dt from Eqs. (345) by the relation

$$dt = \frac{ds}{V}, \quad \ldots \ldots \ldots (351)$$

and we have

$$\left. \begin{array}{l} N \cos \theta_x = MV^2 \dfrac{d^2x}{ds^2} - X; \\[4pt] N \cos \theta_y = MV^2 \dfrac{d^2y}{ds^2} - Y; \\[4pt] N \cos \theta_z = MV^2 \dfrac{d^2z}{ds^2} - Z. \end{array} \right\} \quad \ldots \quad (352)$$

Squaring and adding, we have

$$N^2(\cos^2 \theta_x + \cos^2 \theta_y + \cos^2 \theta_z) = M^2 V^4 \left\{ \left(\frac{d^2x}{ds^2}\right)^2 + \left(\frac{d^2y}{ds^2}\right)^2 + \left(\frac{d^2z}{ds^2}\right)^2 \right\}$$
$$- 2MV^2 \left\{ X\frac{d^2x}{ds^2} + Y\frac{d^2y}{ds^2} + Z\frac{d^2z}{ds^2} \right\}$$
$$+ X^2 + Y^2 + Z^2. \quad \ldots \quad (353)$$

Let ρ be the radius of curvature of the surface at any point, and ϕ the angle which the resultant R makes with it; then substituting in Eq. (353) the following values:

$$\cos^2 \theta_x + \cos^2 \theta_y + \cos^2 \theta_z = 1; \quad \ldots \quad (354)$$

$$X^2 + Y^2 + Z^2 = R^2; \quad \ldots \ldots (355)$$

$$R \sin \phi = M \frac{d^2s}{dt^2} = MV^2 \frac{d^2s}{ds^2}; \quad \ldots \quad (356)$$

$$\left.\begin{aligned}\frac{d\frac{dx}{ds}}{ds} &= \frac{d^2x}{ds^2} - \frac{dx}{ds}\cdot\frac{d^2s}{ds^2};\\ \frac{d\frac{dy}{ds}}{ds} &= \frac{d^2y}{ds^2} - \frac{dy}{ds}\cdot\frac{d^2s}{ds^2};\\ \frac{d\frac{dz}{ds}}{ds} &= \frac{d^2z}{ds^2} - \frac{dz}{ds}\cdot\frac{d^2s}{ds^2};\end{aligned}\right\} \quad \ldots \ldots (357)$$

$$\left\{\frac{d^2x}{ds^2}\right\}^2 + \left\{\frac{d^2y}{ds^2}\right\}^2 + \left\{\frac{d^2z}{ds^2}\right\}^2 = \left\{\frac{d^2s}{ds^2}\right\}^2 + \frac{1}{\rho^2}; \quad (358)$$

the last being obtained from Eq. (7),—we have

$$N^2 = M^2V^4\left\{\left(\frac{d^2s}{ds^2}\right)^2 + \frac{1}{\rho^2}\right\}$$

$$- 2MV^2\left\{X\left(\frac{d\frac{dx}{ds}}{ds} + \frac{dx}{ds}\cdot\frac{d^2s}{ds^2}\right) + Y\left(\frac{d\frac{dy}{ds}}{ds} + \frac{dy}{ds}\cdot\frac{d^2s}{ds^2}\right)\right.$$

$$\left. + Z\left(\frac{d\frac{dz}{ds}}{ds} + \frac{dz}{ds}\cdot\frac{d^2s}{ds^2}\right)\right\} + R^2$$

$$= R^2\sin^2\phi + \frac{M^2V^4}{\rho^2}$$

$$- 2MV^2\left\{\frac{X}{R}\rho\frac{d\frac{dx}{ds}}{ds} + \frac{Y}{R}\rho\frac{d\frac{dy}{ds}}{ds} + \frac{Z}{R}\rho\frac{d\frac{dz}{ds}}{ds}\right\}\frac{R}{\rho}$$

$$- 2MV^2\frac{d^2s}{ds^2}R\left\{\frac{X}{R}\frac{dx}{ds} + \frac{Y}{R}\frac{dy}{ds} + \frac{Z}{R}\frac{dz}{ds}\right\} + R^2\sin^2\phi$$

$$+ R^2\cos^2\phi. \ldots (359)$$

But since

$$\frac{X}{R}\rho\frac{d\frac{dx}{ds}}{ds} + \frac{Y}{R}\rho\frac{d\frac{dy}{ds}}{ds} + \frac{Z}{R}\rho\frac{d\frac{dz}{ds}}{ds} = \cos\phi \quad \ldots \quad (360)$$

and

$$\frac{X}{R}\frac{dx}{ds} + \frac{Y}{R}\frac{dy}{ds} + \frac{Z}{R}\frac{dz}{ds} = \sin\phi, \ldots \quad (361)$$

we have finally

$$\begin{aligned}N' &= \frac{M^2V^4}{\rho^2} - \frac{2MV^2R\cos\phi}{\rho} + R^2\cos^2\phi + 2R^2\sin^2\phi - 2R^2\sin^2\phi \\ &= \frac{M^2V^4}{\rho^2} - \frac{2MV^2R\cos\phi}{\rho} + R^2\cos^2\phi; \ldots \quad (362)\end{aligned}$$

whence

$$N = \frac{MV^2}{\rho} - R\cos\phi \quad \ldots \quad (363)$$

and

$$N = R\cos\phi - \frac{MV^2}{\rho}. \quad \ldots \quad (364)$$

Therefore the normal reaction of the surface at any point is equal to the difference between the normal component of the resultant of the extraneous forces and $\frac{MV^2}{\rho}$.

134. When the body is in motion on the concave side of the surface the value of N is given by Eq. (363), and when on the convex side by Eq. (364). In the first case we see then that when the action-line of R lies outside of the tangent to the curved path of the body the intensity of the normal reaction is equal to the sum of $\frac{MV^2}{\rho}$ and $R\cos\phi$, and when it lies within the tangent, to the excess of $\frac{MV^2}{\mu}$ over $R\cos\phi$. When, in the latter case, $R \cos \phi$ becomes greater than $\dfrac{MV^2}{\rho}$, the body will leave the concave surface and describe a path of greater curvature, since N can never become negative.

In the second case Eq. (364) shows that the body can only remain on the given surface as long as $R \cos \phi$ is positive and numerically greater than $\dfrac{MV^2}{\rho}$; when this condition is not ful-

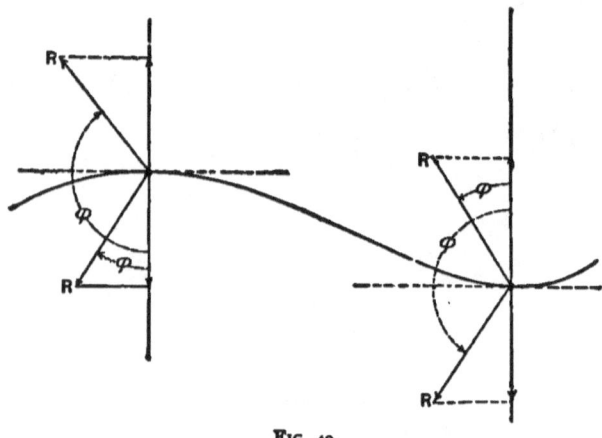

FIG. 49.

filled the body will leave the surface and describe a path of less curvature. Fig. 49 illustrates the two cases.

135. *Centrifugal Force.*—The force whose intensity is measured by $\dfrac{MV^2}{\rho}$ was formerly supposed to be exerted *by the body itself* and to act *from the centre of curvature outward.* It was therefore called *centrifugal force.* This name is still in general use, but it is evidently a misnomer arising from erroneous conclusions from well-known phenomena. We have seen, Art. 16, that when the path is a curve the total acceleration is the resultant of two rectangular components, one, $\dfrac{d^2s}{dt^2}$, in the direction of the tangent, and the other, $\dfrac{v^2}{\rho}$, along the radius of curvature *towards* *the centre*. The intensities of the corresponding forces are therefore $M\dfrac{d^2s}{dt^2}$ and $M\dfrac{V^2}{\rho}$.

In the discussion above, we find the latter force to be that part of the normal reaction of rigid curves or surfaces which is called into play by the equal direct pressure due to the change in the direction of motion of the body. Hence $\dfrac{MV^2}{\rho}$ is the intensity of that force which actually deflects the body from its rectilinear path. It varies directly as the square of the velocity of the body, and inversely as the radius of curvature of its path; it is zero when ρ is ∞, that is, when the path is a right line, and infinite when ρ is zero, that is, no finite force can abruptly change the direction of motion of a body.

Let a body B, Fig. 50, be whirled about a centre by means of a cord dB held by the hand at d, the latter describing the small circumference. The pull on the cord, represented by Ba, has two components, Bb and Bc; the former accelerates the motion of B, while the latter deflects it from the rectilinear path which it would follow due to its acquired velocity and the component Bb. If the motion be accelerated until $\dfrac{MV^2}{\rho}$, equal to $Ba \cos \phi$, is greater than can be applied through the medium of the cord, the latter will break and the body will continue to move in the direction of its motion at the instant of rupture.

Fig. 50.

Due to the great angular velocities with which fly-wheels, grindstones, etc., are often made to rotate, they sometimes break in pieces; this occurs when the cohesion at the sections of rupture is less than the pull required to cause the parts of the body to describe their circular paths with the required velocity. When the areas of section of such bodies and the values of the cohesion per unit of area are known, the safe values of V can readily be determined from the expression $\dfrac{MV^2}{\rho}$.

136. Referring to Art. 44, we can now see why the rotation of the earth on its axis diminishes the weight of bodies, and makes the apparent less than the actual weight. Since the mass m, Fig. 51, is constrained to remain on the convex side of its parallel of latitude, the normal reaction is

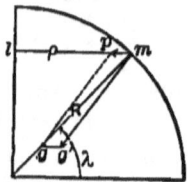

Fig 51.

$$N = mg \cos \lambda - \frac{mv^2}{\rho}$$
$$= \cos \lambda (g - \omega^2 R)m, \quad . \quad . \quad (365)$$

in which ω is the angular velocity of the earth, λ the latitude of m, and ρ and R the radii of the parallel and of the earth at m respectively. At the equator the acceleration due to gravity is diminished by the value of $\omega^2 R$, or

$$\omega^2 R = \frac{4\pi^2 \, 3962.72 \times 5280}{(24 \times 60 \times 60)^2} = 0.1106 \, f. \, s., \quad . \quad . \quad (366)$$

which, as was previously stated, is about $\frac{1}{289}$ that due to gravity; hence if the angular velocity of the earth were seventeen times as great as it now is, bodies at the equator would have no *apparent* weight.

137. *Problems in Constrained Motion.*—To solve these problems we may either make use of the equations of constraint (342), together with the equation of the surface, or by means of Eqs. (345) treat them as cases of free motion. By the first method we find the partial differential coefficients from the equation of the surface and substitute them, with the component intensities of the extraneous forces, in the equations of constraint. We will have then two equations involving three second differential coefficients, and a third equation can be obtained by differentiating the equation of the surface; hence we can thus find a single equation by their combination, involving only a single second differential coefficient, its corresponding variable and constants. If integration be possible, the solution of the problem can then be accomplished. By the second method the substitution of the normal reaction and the component intensities of the extraneous forces in Eqs. (345) gives three equations, each involving the corresponding component acceleration; the steps are then those employed in cases of free motion.

138. *On an Inclined Plane*.—Let the forces be friction and the weight of the body. Assume that the friction is constant and directly opposed to the motion, and let F be its intensity. Let Mg be the weight of the body, and take the axis of y in the inclined plane, the latter making the angle α with the axis of x. Let the body start from rest at the origin; the motion will then be in the plane xz, and we have for the equation of the path

$$L = z + x \tan \alpha = 0. \quad \ldots \ldots \quad (367)$$

Assume the second of Eqs. (342), and substitute in it the following values:

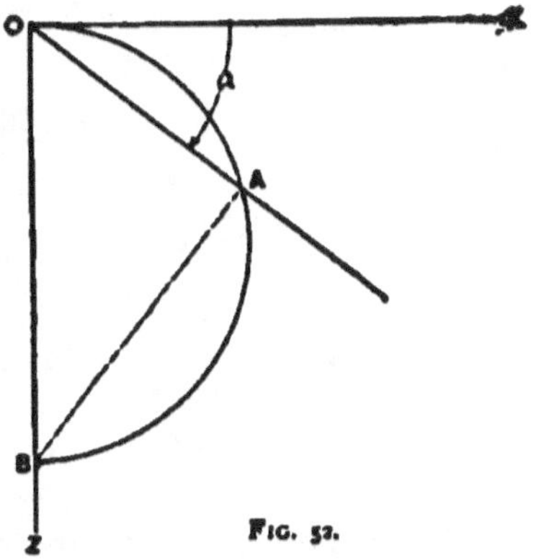

$$\left. \begin{aligned} \frac{dL}{dz} &= 1; \\ \frac{dL}{dx} &= \tan \alpha; \\ d^2x &= -\frac{d^2z}{\tan \alpha}; \\ X &= -F \cos \alpha; \\ Z &= -Mg + F \sin \alpha. \end{aligned} \right\} \quad (368)$$

This gives us

$$\left. \begin{aligned} \frac{d^2z}{dt^2} &= \frac{F \sin \alpha + F \sin \alpha \tan^2 \alpha - Mg \tan^2 \alpha}{M(1 + \tan^2 \alpha)} \\ &= \frac{1}{M}\left(F \sin \alpha - Mg \frac{\tan^2 \alpha}{\sec^2 \alpha}\right) = \sin \alpha \left(\frac{F}{M} - g \sin \alpha\right). \end{aligned} \right\} (369)$$

FIG. 52.

Multiplying by $2dz$ and integrating, we get

$$\frac{dz^2}{dt^2} = (V_z)^2 = 2 \sin \alpha \left(\frac{F}{M} - g \sin \alpha\right)z, \quad \ldots \quad (370)$$

or, solving with respect to dt,

$$dt = \frac{1}{\sqrt{2 \sin \alpha \left(\frac{F}{M} - g \sin \alpha\right)}} \frac{dz}{\sqrt{z}}; \quad \ldots \quad (371)$$

and by integration,

$$z = \frac{1}{2} \sin \alpha \left(\frac{F}{M} - g \sin \alpha\right) t^2. \quad \ldots \quad (372)$$

If $F = 0$, we have

$$z = -\tfrac{1}{2} g t^2 \sin^2 \alpha \quad \ldots \quad (373)$$

and

$$V_z = \sin \alpha \sqrt{-2gz}. \quad \ldots \quad (374)$$

We also readily get

$$V_x = \frac{V_z}{\tan \alpha} = \cos \alpha \sqrt{-2gz}; \quad \ldots \quad (375)$$

$$V = \sqrt{(V_z)^2 + (V_x)^2} = \sqrt{-2gz} = gt \sin \alpha. \quad (376)$$

Hence the velocity at any point is that due to the height, and varies directly as the product of the time and the sine of the inclination.

Comparing this value of V with that given in the discussion of motion due to gravity alone (Art. 92), we see that V is the same function of z, but a different function of t.

CONSTRAINED MOTION. 155

These results can be more readily obtained by considering the motion to be free, the body being acted upon by the resultant of all the forces, including the normal reaction. This resultant is $Mg \sin \alpha - F$, and it acts in the direction of the path.

139. Let the circumference OAB (Fig. 52) be in the plane xz, and tangent to the axis of x at O. Then we have for the point A, Eq. (373),

$$-z = \tfrac{1}{2}gt^2 \sin^2 \alpha = OA \sin \alpha = OB \sin^2 \alpha \quad . \quad . \quad (377)$$

or

$$t = \sqrt{\frac{2OB}{g}}, \quad . \quad . \quad . \quad . \quad . \quad (378)$$

which is the time of fall from rest down the diameter of the circle.

Hence the time required for the body to pass over any chord of a given circle, the plane of the circle being vertical and the body starting from the upper extremity of the vertical diameter, is independent of its length; that is, the circumference is the locus of simultaneous arrival down all right lines in a vertical plane, these lines having a common point at the origin of motion. The circle is called the *synchronous curve* of such lines.

From this property of the circle the right line of quickest descent from a given point to a given right line in a vertical plane containing the point, or from the given right line to the given point, can be found. In the first case let P, Fig. 53, be the point and AB the line; draw the horizontal PA meeting the line at A, and bisect the angle PAB; the point of meeting C of the bisector and the vertical line through P is the centre of the circle whose chord PO is the required line.

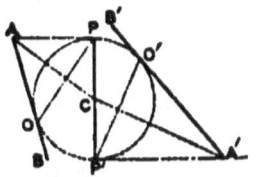

Fig. 53.

In the second case let $B'A'$ be the right line and P' the point; draw the horizontal

$P'A'$ meeting $B'A'$ at A'; bisect the angle $P'A'B'$, and the chord $O'P'$ of the circle $PO'P'$ is the required line.

140. On the Concave Side of the Arc of a Cycloid whose Plane is Vertical.—Take the origin as in Fig. 54. The equation of the cycloid is

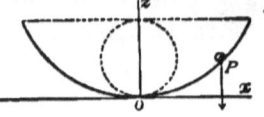

Fig. 54.

$$x = a \text{ versin}^{-1} \frac{z}{a} + (2az - z^2)^{\frac{1}{2}}. \quad (379)$$

Considering the weight as the only force, we have $X = 0$, $Z = -Mg$, and Eqs. (345) become

$$\left. \begin{array}{l} \dfrac{d^2x}{dt^2} + \dfrac{N}{M} \dfrac{dz}{ds} = 0, \\[6pt] \dfrac{d^2z}{dt^2} + g - \dfrac{N}{M} \dfrac{dx}{ds} = 0, \end{array} \right\} \quad \cdots \quad (380)$$

since

$$\cos \theta_x = -\frac{dz}{ds} \quad \text{and} \quad \cos \theta_z = \frac{dx}{ds}.$$

Multiplying by $2dx$ and $2dz$, respectively, and adding the products, we obtain

$$\frac{2dx\,d^2x + 2dz\,d^2z}{dt^2} + 2g\,dz = 0.$$

Integrating, we have

$$\frac{dx^2 + dz^2}{dt^2} = \frac{ds^2}{dt^2} = V^2 = -2gz + C. \quad \cdots \quad (381)$$

Supposing V to be zero when $z = h$, this reduces to

$$\frac{ds^2}{dt^2} = V^2 = 2g(h-z); \quad \ldots \ldots \quad (382)$$

or, the velocity is that due to the vertical distance through which the centre of mass has fallen.

The normal reaction is, Eq. (363),

$$N = \frac{MV^2}{\rho} - R \cos \phi = \frac{MV^2}{\rho} + Mg\frac{dx}{ds}. \quad \ldots \quad (383)$$

By differentiating the equation of the curve, we get

$$\frac{dx}{(2a-z)^{\frac{1}{2}}} = \frac{dz}{z^{\frac{1}{2}}} = \frac{ds}{(2a)^{\frac{1}{2}}}; \quad \ldots \ldots \quad (384)$$

and since the radius of curvature is

$$\rho = [8a(2a-z)]^{\frac{1}{2}}, \quad \ldots \ldots \quad (385)$$

these values give, for N,

$$N = \frac{MV^2}{[8a(2a-z)]^{\frac{1}{2}}} + Mg\frac{(2a-z)^{\frac{1}{2}}}{(2a)^{\frac{1}{2}}} = Mg\frac{(h+2a-2z)}{[2a(2a-z)]^{\frac{1}{2}}}. \quad (386)$$

If the body start from rest at the highest point of the curve, then will $V^2 = 2gh = 4ag$ when it reaches the lowest point; then $z = 0$, and we have

$$N = Mg + Mg = 2Mg, \quad \ldots \ldots \quad (387)$$

or double the weight; therefore the direct pressure *due to the velocity* generated by the weight in falling from the highest to the lowest point is equal to the weight of the body.

Let the body start from any point, as P; then $OP = s$, and we have, Eqs. (382) and (384),

$$t = -\frac{1}{(2g)^{\frac{1}{2}}}\int_h^0 \frac{ds}{(h-z)^{\frac{1}{2}}} = \left(\frac{a}{g}\right)^{\frac{1}{2}} \int_0^h \frac{dz}{(hz-z^2)^{\frac{1}{2}}}$$
$$= \left(\frac{a}{g}\right)^{\frac{1}{2}} \left\{ \text{versin}^{-1} \frac{2z}{h} \right\}_0^h = \pi \left(\frac{a}{g}\right)^{\frac{1}{2}}; \quad \ldots \ldots \quad (388)$$

that is, the time of descent from any point on the curve to the lowest point is independent of its height h and will be the same no matter from what point the body starts. The velocity with which it passes the lowest point will depend on the vertical height of fall, and is equal to $\sqrt{2gh}$: due to this velocity the body will ascend to an equal height on the other branch of the cycloid in a time equal to that of the descent, or $\pi\left(\frac{a}{g}\right)^{\frac{1}{2}}$; after which it will return to the point of starting, and so on continuously. These recurring movements are called *oscillations*, and since they are performed in equal times, $2\pi\left(\frac{a}{g}\right)^{\frac{1}{2}}$, the cycloid is called a *tautochronous curve*.

The time down any inclined right line l is, Eq. (378),

$$t = \left(\frac{2l}{g \sin \alpha}\right)^{\frac{1}{2}}; \quad \ldots \ldots \quad (389)$$

therefore down any radius of curvature of the cycloid it is

$$t = \left\{ \frac{2\rho}{g \frac{dx}{ds}} \right\}^{\frac{1}{2}}; \quad \ldots \ldots \quad (390)$$

whence, substituting the value of ρ, Eq. (385), and of $\frac{dx}{ds}$, Eq.

(384), and reducing, we get

$$t = \left(\frac{8a}{g}\right)^{\frac{1}{2}}, \quad \ldots \ldots \ldots \quad (391)$$

which is the time of fall down the maximum radius of curvature, or twice the diameter of the generating circle. Therefore the times of descent down all radii of curvature of the cycloid are equal.

141. *On the Concave Side of a Circular Arc in a Vertical Plane.* —Let a be the radius, and take the origin at the highest point with the axis of z vertical and positive downward. The equation of the circle is

$$x^2 = 2az - z^2, \quad \ldots \ldots \ldots \quad (392)$$

from which we have

$$\frac{dx}{a-z} = \frac{dz}{x} = \frac{ds}{a}. \quad \ldots \ldots \quad (393)$$

Let the weight be the only force acting, and denote the velocity at the origin by V_0. Then we have

$$MV^2 - MV_0^2 = M(2gz), \quad \ldots \ldots \quad (394)$$

or

$$V^2 = 2gz + V_0^2. \quad \ldots \ldots \ldots \quad (395)$$

For the normal reaction we have

$$N = \frac{MV^2}{a} - Mg\frac{dx}{ds} = M\frac{2gz + V_0^2}{a} - Mg\frac{a-z}{a}. \quad (396)$$

The limit of N being zero, we have for the corresponding value of the velocity at the origin

$$V_0 = \sqrt{ag}. \quad \ldots \ldots \ldots \quad (397)$$

That is, the body will not follow the curve continuously in one direction unless the velocity at the highest point be at least equal to $\sqrt{ag}$. The corresponding value of the velocity at the lowest point is $\sqrt{5ag}$.

Constrained Motion about a Fixed Axis.

142. *The Compound Pendulum.*—When a rigid solid oscillates freely about a horizontal axis under the action of its weight, it is called a compound pendulum. Let G, Fig. 55, be the centre of gravity, h its distance from the axis, and ψ the angle which h makes with the vertical plane through the axis. Then we have, Eq. (283),

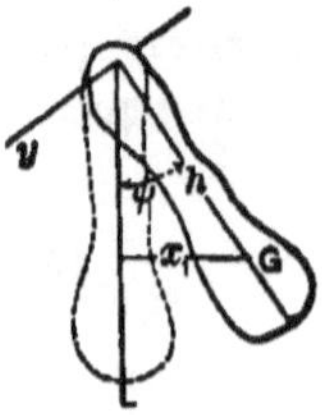

FIG. 55.

$$\frac{d^2\psi}{dt^2} = \frac{M_t}{\Sigma mr^2} = -\frac{Mgh \sin \psi}{M(k_t^2 + h^2)}, \quad \cdots \quad (398)$$

or, taking ψ so small that it may be substituted for its sine,

$$\frac{d^2\psi}{dt^2} = -\frac{gh\psi}{k_t^2 + h^2}. \quad \cdots \quad \cdots \quad (399)$$

Multiplying by $2d\psi$ and integrating, supposing the body to start from rest when $\psi = \alpha$, we have

$$\frac{d\psi^2}{dt^2} = \frac{gh}{k_t^2 + h^2}(\alpha^2 - \psi^2); \quad \cdots \quad \cdots \quad (400)$$

and

$$dt = \sqrt{\frac{k_t^2 + h^2}{gh}} \frac{-d\psi}{\sqrt{\alpha^2 - \psi^2}}. \quad \cdots \quad (401)$$

Hence

$$t = \sqrt{\frac{k_i^2 + h^2}{gh}} \int \frac{-d\psi}{\sqrt{\alpha^2 - \psi^2}} = \sqrt{\frac{k_i^2 + h^2}{gh}} \int \frac{-\frac{d\psi}{\alpha}}{\sqrt{1 - \frac{\psi^2}{\alpha^2}}}$$

$$= \sqrt{\frac{k_i^2 + h^2}{gh}} \cos^{-1}\frac{\psi}{\alpha} + C. \quad (402)$$

The time of one oscillation is therefore

$$t_i = \pi\sqrt{\frac{k_i^2 + h^2}{gh}}, \quad \ldots \quad (403)$$

which is the integral between the limits $\psi = \alpha$ and $\psi = -\alpha$.

The oscillations of a compound pendulum may therefore be considered isochronal when the arcs of vibration are very small.

143. *The Equivalent Simple Pendulum.* — If in Eq. (403) we make $k_i = 0$, we shall have the time of oscillation of a material point about a horizontal axis with which it is connected by a line without weight. Such a pendulum is called a *simple pendulum*. Denoting its length by l, we have for its time of oscillation

$$t' = \pi\sqrt{\frac{l}{g}}. \quad \ldots \quad (404)$$

For a simple and a compound pendulum which are isochronal with each other we have

$$t_i = t',$$

or

$$l = \frac{k_i^2 + h^2}{h}; \quad \ldots \quad (405)$$

and l is called the *equivalent simple pendulum* of the given compound pendulum.

A point of the compound pendulum on the line h at the distance l from the axis is called the *centre of oscillation*, and a line through this centre and parallel to the axis of suspension is called the *axis of oscillation*.

144. From Eq. (405) we have

$$(l - h)h = k_i^2. \quad \ldots \ldots \ldots (406)$$

Hence, as regards their distances from the centre of gravity, the axes of suspension and oscillation are connected by the same law as the line of impact and spontaneous axis. Therefore *the axes of suspension and oscillation are reciprocal and convertible, and the times of oscillation about them are the same.*

145. Adding $2k_i$ to both members of Eq. (405), we have, after reduction,

$$l = 2k_i + \frac{(h - k_i)^2}{h}. \quad \ldots \ldots (407)$$

The minimum value of l is $2k_i$, and this occurs when $h = k_i = l - h$. Therefore, since t is a minimum when l is a minimum, *the time of oscillation is a minimum when the axis of suspension passes through the centre of gyration with respect to an axis through the centre of gravity and parallel to the axis of suspension.* It is evidently a minimum minimum for that centre of gyration which corresponds to the least central principal moment of inertia of the body.

146. *The Simple Seconds Pendulum.*—The *simple seconds pendulum* is a simple pendulum whose time of oscillation is one mean solar second. Its length, $L = \frac{g}{\pi^2}$, obtained by making $t = 1$ in Eq. (404), varies directly with the acceleration due to gravity.

Let l be the equivalent simple pendulum of a given compound pendulum, and we have

$$t^2 = \pi^2 \frac{l}{g}; \quad \ldots \ldots \ldots (408)$$

and substituting in the value of L, we get

$$L = \frac{g}{\pi^2} = \frac{l}{t^2}. \quad \ldots \ldots \ldots \quad (409)$$

Hence to find L it is necessary simply to find the time of oscillation of a compound pendulum and the length of its equivalent simple pendulum. To do this Kater used the pendulum represented in Fig 56. Its centre of gravity is removed from its middle point by the heavy bob BB'; S and O are two knife-edged prismatic axes of hardened steel, permanently attached to the rod and having their edges turned towards each other; m is a small ring-shaped mass which can be moved up and down so as to change slightly the position of G and the value of k_g, and is arranged with clamps and a screw to fix it in any desired position. For any assumed position of m let GC be the gyratory circumference which, from the construction of the pendulum, always lies between S and G, and beyond O from G, as in the figure.

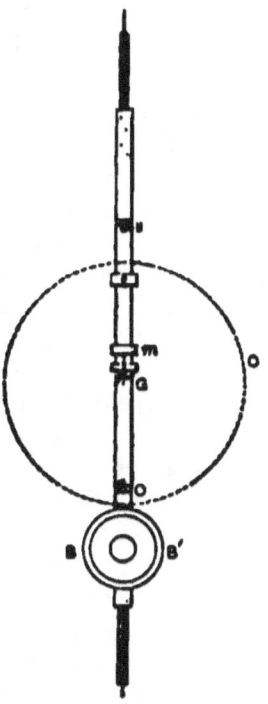

Fig. 56.

It is evident that with this arrangement every position of m gives a different compound pendulum for each axis, and that there is but one among these whose equivalent simple pendulum is the length $h+h'$ between the fixed axes, and for which the times of oscillation about S and O would be the same. It was this particular compound pendulum that Kater desired to find experimentally by moving the sliding mass m until it was placed in the required position. This he was enabled to do by the application of the principles stated in Arts. 144 and 145.

By the first principle, the times of oscillation about S and O

are equal only when these lines accurately coincide with the axes of suspension and oscillation. On trial the times about S and O would in general be found to differ slightly, and by the second principle a displacement of m towards either axis would cause both times to lengthen or both to shorten, *but unequally*. And since the gyratory circumference is nearer O than S, a greater change for the same displacement takes place with respect to the latter than to the former, from the principle of maxima and minima. The pendulum was mounted on the axis S and permitted to oscillate through small arcs, the number of oscillations being counted; then it was suspended from the axis O and oscillated, the corresponding number in the same time being determined. Repeated trials enabled Kater to find the position of m for which the distance between S and O, or $h + h' = l$, was accurately the length of the equivalent simple pendulum, and its time of oscillation was known from observation. This method does not require the determination of the exact position of G, but has the disadvantage of exacting accurate adjustment of the mass m, an operation requiring very careful and repeated manipulation.

In order to count the number of oscillations the method of coincidences was used. Thus the pendulum was mounted in front of a clock whose pendulum, beating seconds, could be seen, by means of a telescope, behind the position of the Kater pendulum, when it passed the lowest point of its arc of oscillation. At a certain second, indicated by the clock, the two pendulums would coincide, and after an exact number of oscillations of the clock pendulum they would again coincide. The number of oscillations of the clock pendulum in this period, if the duration of the clock pendulum's oscillation was *less* than that of the Kater pendulum, would be *two more*, and if *greater two less*, than the Kater pendulum. The clock gives its own indication, and hence the other is at once determined. In Kater's experiment the entire duration of each trial lasted about thirty-five minutes, corresponding to five coincidences, or four intervals of 530 seconds each.

147. Another and less tedious method is to use a reversible pendulum of such a form that its centre of gravity may be determined with considerable accuracy, and whose axes are not reciprocal. Then we have, Eq. (403),

$$\left.\begin{array}{l} k_i^2 + h^2 = Lht^2; \\ k_i^2 + h'^2 = Lh't'^2; \end{array}\right\} \quad \ldots \quad (410)$$

and eliminating k_i,

$$\left.\begin{array}{l} \dfrac{h^2 - h'^2}{L} = ht^2 - h't'^2; \\ \dfrac{h + h'}{L} = \dfrac{ht^2 - h't'^2}{h - h'} = \dfrac{t^2 + t'^2}{2} + \dfrac{1}{2}\dfrac{h + h'}{h - h'}(t^2 - t'^2); \\ \dfrac{1}{L} = \dfrac{1}{2}\dfrac{t^2 + t'^2}{h + h'} + \dfrac{1}{2}\dfrac{t^2 - t'^2}{h - h'}. \end{array}\right\} \quad (411)$$

As the error in this method is due to the approximate values of h and h', t and t' should be as nearly equal as practicable, and h and h' should differ as much as practicable, thus making the term which depends on $h - h'$ very small.

148. *The Value of g.*—Having found L by experiment, the acceleration due to gravity is found from

$$g = \pi^2 L. \quad \ldots \quad \ldots \quad (412)$$

When g has been determined for one locality we may find its value for any other place by means of any compound pendulum. For the two places we have

$$t = \pi\sqrt{\dfrac{k_i^2 + h^2}{gh}} \quad \text{and} \quad t' = \pi\sqrt{\dfrac{k_i^2 + h^2}{g'h}}. \quad \ldots \quad (413)$$

Hence

$$t^2 : t'^2 :: g' : g, \quad \ldots \ldots \quad (414)$$

or

$$g' = \frac{t^2}{t'^2} g. \quad \ldots \ldots \quad (415)$$

If N and N' be the numbers of oscillations per hour at the two places, these become

$$N'^2 : N^2 :: g' : g; \quad \ldots \ldots \quad (416)$$

$$g' = \frac{N'^2}{N^2} g. \quad \ldots \ldots \quad (417)$$

The formula Eq. (59), taken from Everett's Units and Physical Constants, gives, for all latitudes,

$$g = 32.173 - 0.0821 \cos 2\lambda - .000003h, \quad \ldots \quad (418)$$

and from $L = \dfrac{g}{\pi^2}$ we have the corresponding value for the simple seconds pendulum,

$$L = 3.2597 - .0083 \cos 2\lambda - .0000003h. \quad \ldots \quad (419)$$

Substituting the value of the latitude of West Point, $41°23'31''$, we derive

$$g = 32.163 f.s. \quad \text{and} \quad L = 3.2587 ft. \quad \ldots \quad (420)$$

for the values of g and L at West Point.

149. *Length of the Equivalent Simple Pendulum.*—The length of the equivalent simple pendulum of any compound pendulum is, Eq. (409),

$$l = Lt^2, \quad \ldots \ldots \quad (421)$$

in terms of its time of oscillation and the length of the simple seconds pendulum at the place of observation.

150. *British Standard of Length.*—In 1824 an act of Parliament defined the *Imperial Standard Yard* "to be the straight line or distance between two points in the gold studs in the straight brass rod" known as the "Standard Yard, 1760," at 62° F., and designated the ratio of its length to that of a simple pendulum vibrating mean seconds, in vacuo, at sea-level at the latitude of London, to be as 36 : 39.1393. This standard was destroyed in the burning of the Houses of Parliament in 1834. Upon the recommendation of a commission of scientific men appointed to restore the standards of weights and measures, the act of 1855 defined the *Imperial Standard Yard* of Great Britain to be "the straight line or distance between the centres of the two gold plugs or pins in the bronze bar deposited in the office of the Exchequer," at 62° F. Its restoration in case of loss or destruction is provided for, by reference to its numerous copies. The present standard is therefore not referred to the length of the simple seconds pendulum.

151. *To determine the Moment of Inertia of a body by the principles of the Compound Pendulum.*—It is often necessary to find the moment of inertia of a body which is not homogeneous nor of a regular form, and to which therefore the methods of Art. 101 will not apply. Whenever the body can be mounted so as to oscillate about a horizontal axis under the action of its own weight, we may apply the principles of the compound pendulum and find its moment of inertia no matter what its form or substance may be

Thus, multiplying Eq. (405) by M, the mass of the body, and clearing of fractions we have

$$Mlh = M(k_1^2 + h^2) = \Sigma mr^2; \quad \ldots \quad (422)$$

that is, the product of the *mass of the body*, the *length of its equivalent simple pendulum*, and the *distance of the axis from the centre of gravity*, is the moment of inertia with respect to the axis about which the body is oscillated. The first two quantities are obtained from the equations

$$M = \frac{W}{g} \quad \text{and} \quad l = Lt^2, \quad \ldots \quad (423)$$

in which W is the weight in pounds, g the acceleration due to gravity at the place of observation, derived from Eq. (418), L the length of the simple seconds pendulum, and t the time of oscillation of the body about the axis in question. The value of h can be found by measurement, provided the exact position of the centre of gravity is known. When this is not known h may be found as follows: Attach a dynamometer to the extremity farthest from the axis of suspension, as in Fig. 57, and take its reading when the body has been lifted by the dynamometer to a position such that the axis and centre of gravity are in a horizontal plane; this position is reached when the reading of the dynamometer is a maximum. Then if R be the reading in pounds and a the horizontal distance TS in feet, we have, by the equality of moments,

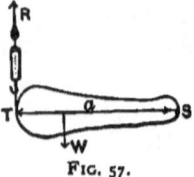

FIG. 57.

$$Ra = Wh,$$

whence

$$h = \frac{Ra}{W}. \quad \ldots \quad \ldots \quad (424)$$

Substituting these values of M, l and h, we have

$$Mlh = \Sigma mr^2 = \frac{W}{g} \cdot \frac{Ra}{W} \cdot Lt^2 = \frac{Ra}{\pi^2} t^2. \quad \ldots \quad (425)$$

To find the moment of inertia with respect to an axis through the centre of mass, the body must be mounted on a parallel axis, with reference to which its moment of inertia may be found by the above method. Its moment of inertia with respect to the axis through the centre of mass is then

$$Mk_1^2 = Mlh - Mh^2. \quad \ldots \ldots \quad (426)$$

For bodies of small mass it is sometimes more convenient to attach them to a pendulum whose moment of inertia with respect to its axis is known. The moment of inertia of the combination may then be determined, and the difference between this result and the moment of inertia of the known pendulum is the moment of inertia of the body with respect to the axis of the pendulum, from which the required moment may be found.

152. *The Conical Pendulum.*—Let a simple pendulum have a component vibration about two horizontal axes at right angles to each other. The path of the material point will in general be a curve of double curvature on the surface of a sphere whose radius is the length of the pendulum, this length describing a cone with its vertex at the point of suspension. Such a pendulum is called a *conical pendulum*.

The equation of the projection on a horizontal plane of the curve of double curvature may be determined as follows:

Let ψ and ϕ be the arcs of oscillation about the two axes respectively; then $\sin \psi$ and $\sin \phi$ will be the co-ordinates of the projected path of the material point of the pendulum. If the arcs be taken so small that the oscillations may be considered isochronal, then ψ and ϕ may be taken as co-ordinates of the projected path, instead of $\sin \psi$ and $\sin \phi$.

Let β and α be the maximum values of ϕ and ψ respectively. When $\phi = \beta$ we may take $\psi = 0$, and $\phi = 0$ when $\psi = \alpha$. Taking $\psi = \alpha$ when $t = 0$, and taking $\sqrt{\dfrac{l}{g}}$ seconds as the unit in which t is measured, we have, Eq. (402),

$$t = \cos^{-1} \frac{\psi}{\alpha}; \quad \ldots \ldots \quad (427)$$

and taking $\phi = \beta$ when $\psi = 0$, or $t = \tfrac{1}{2}\pi$, we have

$$t = \cos^{-1}\frac{\phi}{\beta} + \tfrac{1}{2}\pi. \quad \ldots \quad \ldots \quad (428)$$

Eliminating t, we get

$$\cos^{-1}\frac{\psi}{\alpha} = \cos^{-1}\frac{\phi}{\beta} + \tfrac{1}{2}\pi; \quad \ldots \quad (429)$$

and taking the cosines of both members,

$$\frac{\psi}{\alpha} = -\sqrt{1 - \frac{\phi^2}{\beta^2}}, \quad \ldots \quad (430)$$

or

$$\psi^2\beta^2 + \phi^2\alpha^2 = \alpha^2\beta^2, \quad \ldots \quad (431)$$

the equation of an ellipse referred to its centre and axes.

153. While the point moves in azimuth about the vertical through the point of suspension, the change in azimuth from the time when $\psi = \alpha$ until $\phi = \beta$ is $\dfrac{\pi}{2}$. But this deduction is made under the supposition that the vibrations of a simple pendulum are isochronal. However, as the time of vibration increases with the length of the arc, the shorter component vibration in the conical pendulum will be completed first, counting the time from any assumed epoch. Therefore the change in azimuth from the time when $\psi = \alpha$ until $\psi = -\alpha$ is greater than π, and that from $\psi = \alpha$ until $\phi = \beta$ is greater than $\dfrac{\pi}{2}$. Hence the axis of the ellipse has a motion in azimuth in the same direction as that of the pendulum.

A closer approximation shows that the change in azimuth

from $\phi = \alpha$ until $\phi = \beta$ is $\frac{\pi}{2}\left(1 + \frac{3}{8}\alpha\beta\right)$ instead of $\frac{\pi}{2}$. The ellipse therefore makes one complete revolution in azimuth in $\frac{8}{3\alpha\beta}$ times its periodic time.

Equilibrium.

154. When a body is in a state of equilibrium, Art. 68, the acceleration factors in the general equation of energy become separately equal to zero, and there can then be no change of potential into kinetic energy, or the reverse. Therefore during equilibrium the general equation of energy reduces to

$$\Sigma Idp = 0. \qquad \ldots \ldots \ldots \text{(S)}$$

In general a free body is never in a state of equilibrium, since it is subjected to the action of forces the resultant of which is not in general zero. Hence a body in equilibrium must be under constraint. But since any case of constrained motion may be discussed as one of free motion by introducing the normal reaction of the constraining curve, the general equation of energy may thus be made to apply to all cases of equilibrium.

155. There are three cases in which the acceleration may become zero, viz.:

(1) When the resultant is zero, the body being at rest or having uniform motion.

(2) When the resultant of the system of forces reverses its direction as it passes through zero, thus changing the sign of the resultant acceleration.

(3) When the resultant becomes zero but does not pass through it; in this case there is no change of sign of the resultant acceleration.

First When the body is at rest the forces are called *stresses*, and they produce changes of form and volume, these effects usually being known as *strains;* their investigation properly belongs to Applied Mechanics.

When the accelerations become zero because of the uniform motion of the body, Eq. (S) simply asserts that the quantity of work done positively by some of the forces is equal to that done negatively by the others, the whole quantity of energy added to the system being zero. If a body be supposed at rest at any point of that portion of its path over which it has uniform motion, it will evidently remain there if subjected only to the system of forces which caused it to follow that path. In such a case the body is said to be in *neutral* or *indifferent* equilibrium.

156. To investigate the *second* case let us resume Eq. (132). The forces of gravitation, electricity, etc., or what are known generally as *forces of nature*, are taken to be constant or to vary as some function of the distance; and therefore Eq. (132) is applicable to a body subjected to their action, the normal reaction of the curve on which the body moves being considered as one of the extraneous forces.

Assuming consecutive values of the kinetic energy of the body, we have, after developing the difference of the corresponding states of the function by Taylor's theorem,

$$\tfrac{1}{2}MV_2^2 - \tfrac{1}{2}MV_1^2 = F(x_1 + dx_1,\ y_1 + dy_1,\ z_1 + dz_1) - F(x_1, y_1, z_1)$$
$$= Xdx_1 + Ydy_1 + Zdz_1 \pm (Adx_1^2 + Bdy_1^2 + Cdz_1^2) + \text{etc.} \quad (432)$$

If $\tfrac{1}{2}MV_1^2$ be a maximum or a minimum, we shall have as a condition

$$Xdx_1 + Ydy_1 + Zdz_1 = 0; \quad \ldots \ldots \quad (433)$$

that is, the body will be in equilibrium when it reaches a position where it has a maximum or minimum kinetic energy.

Let this condition be fulfilled, and we have

$$\tfrac{1}{2}MV_2^2 - \tfrac{1}{2}MV_1^2 = \pm (Adx_1^2 + Bdy_1^2 + Cdz_1^2) + \text{etc.} \quad (434)$$

If $\tfrac{1}{2}MV_1^2$ be a maximum the second member of this equation will have the negative sign, and whatever be the value of V_1, it must be greater than V_2; and if the body be slightly displaced from its position of equilibrium and then move from rest under the action of the given system of forces, the direction of the resultant must be such as to bring it back again. In this case the body would oscillate to and fro through its position of equilibrium, and could never depart far from it; the equilibrium of a body is therefore said to be *stable* when it occupies a position corresponding to a maximum value of its kinetic energy.

If $\tfrac{1}{2}MV_1^2$ be a minimum the second member of Eq. (434) will be positive, and hence V_2 must be greater than V_1; and if the body move from rest and from a point very near that corresponding to the minimum value of V, the resultant of the system will act in such a direction as to move it away from its position of equilibrium, to which it would never return. The equilibrium is therefore said to be *unstable* when it occupies a position corresponding to a minimum value of its kinetic energy.

157. In the *third* case, since a function is not necessarily a maximum or a minimum when its differential coefficient is equal to zero, it is evident that cases may arise in which a body will be in equilibrium when its kinetic energy is neither a maximum nor a minimum. For example, a body is in equilibrium as it passes a point of inflection at which the resultant is normal to the path. In this case the equilibrium is stable in one direction and unstable in the other.

158. When a *free* body passes a point in its path corresponding to a maximum or a minimum value of its kinetic energy, this point would be a position of stable or unstable equilibrium, if we suppose the body to be moving on a rigid curve coincident with its path.

Let us take the general case of a body acted upon by its own weight, and subjected to any condition of constraint whatever. Since this condition can have no influence on the velocity, the change of kinetic energy between any two points of the path will be that due to the weight acting over the vertical distance through which the body moves; hence we may write

$$\tfrac{1}{2}MV_2^2 - \tfrac{1}{2}MV_1^2 = Mg(z_2 - z_1), \quad \ldots \quad (435)$$

z being taken vertical and positive downward. From this we see that if the body be in stable equilibrium its centre of mass must occupy a point in its path which is lower than the consecutive points on either side; and similarly if it be in unstable equilibrium it must be at a point which is higher than the consecutive points.

Thus, a pendulum is in stable equilibrium when its centre of gravity occupies the lowest point of its possible path, and in unstable equilibrium when it is at the highest point it can reach under the given conditions of constraint. A homogeneous ellipsoid of three unequal axes, resting on a horizontal plane, has two positions of unstable and one of stable equilibrium; an oblate spheroid has one of stable and many of unstable equilibrium; a prolate spheroid has one of unstable and many of stable equilibrium, and a sphere is an example of indifferent equilibrium. If the centre of gravity of the sphere be not coincident with the centre of figure, it will be in stable equilibrium when the centre of gravity is at the lowest point it can reach, and unstable when at the highest.

Examples. 1. A particle on the concave surface of a sphere is acted on by its weight and by a repulsion from the lowest point of the surface, the latter varying inversely as the square of the distance. Find the position of rest.

Take the origin at the lowest point and the axis of z vertical and positive upward; let r be the distance of the point of rest from the origin, a the radius of the surface, and μ the intensity of the repulsive force at the distance unity; then we have the equation of the surface,

$$x^2 + y^2 + z^2 - 2az = 0;$$

the intensity of the repulsion at the distance r,

$$\frac{\mu}{r^3} = \frac{\mu}{2az};$$

and Eqs. (T$_m$),

$$\left.\begin{aligned} X &= \frac{\mu}{2az}\cdot\frac{x}{r} - N\frac{x}{a} = 0; \\ Y &= \frac{\mu}{2az}\cdot\frac{y}{r} - N\frac{y}{a} = 0; \\ Z &= \frac{\mu}{2az}\cdot\frac{z}{r} - N\frac{z-a}{a} - w = 0; \end{aligned}\right\} \quad \cdot\cdot\;(436)$$

from which we have, substituting in the last of Eqs. (436) the value of N obtained from one of the others and reducing,

$$r^3 = \frac{\mu a}{w} \quad \text{and} \quad z = \frac{1}{2}\sqrt[3]{\frac{\mu^2}{aw^2}}; \quad \cdot\;\cdot\;\cdot\;\cdot\;(437)$$

that is, the particle will remain at rest at any point in the circumference of a horizontal circle whose plane is at the distance given by Eq. (437) above the lowest point of the surface.

If another repellent force whose intensity at a unit's distance is μ' be supposed to act on the particle, the value for r' would be $r'^3 = \frac{\mu' a}{w}$; hence the ratio of the intensities of these forces at the distance unity is given by $\frac{\mu'}{\mu} = \frac{r'^3}{r^3}$, or their intensities are directly as the cubes of the distances at which a heavy particle would remain at rest on the surface of a sphere due to their action. The quadrant electroscope, consisting of a light pith ball joined to a point by a thread, measures the relative intensities of strong electrical charges by the principles of this problem.

2. To find the position of rest of a heavy particle m on a given rigid curve AB, Fig. 58, when acted on by its weight w and a constant attraction t toward the origin. Let O be the origin, Ox and Oz the co-ordinate axes, the latter being vertical and positive downwards, and m the supposed position of rest of the particle; let $OA = a$, $Om = r$, and $mOn = \theta$. Then we have

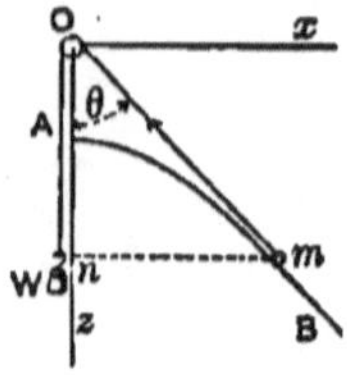

FIG. 58.

$$\left.\begin{aligned} X &= -t\sin\theta + N\frac{dz}{ds} = 0; \\ Z &= w - t\cos\theta - N\frac{dx}{ds} = 0; \end{aligned}\right\} \quad \cdots \quad (438)$$

from which we get

$$(w - t\cos\theta)dz - t\sin\theta dx = wdz - t\frac{xdx + zdz}{r} = 0. \quad (439)$$

But

$$xdx + zdz = rdr, \quad \cdots \cdots \quad (440)$$

and Eq. (439) reduces to

$$wdz - tdr = 0, \quad \cdots \cdots \quad (441)$$

which is the condition of equilibrium. In order that there may be a position of rest on the curve this condition must be satisfied by the co-ordinates of one of its points.

Let the curve be a hyperbola, and we have

$$b^2 z^2 - a^2 x^2 = a^2 b^2; \quad \cdots \cdots \quad (442)$$
$$r^2 = x^2 + z^2 = e^2 z^2 - b^2; \quad \cdots \cdots \quad (443)$$
$$rdr = e^2 z dz; \quad \cdots \cdots \quad (444)$$

and Eq. (441) becomes

$$rwdz - te^2sdz = 0. \quad \ldots \quad (445)$$

Solving Eq. (443) with respect to z and substituting in the resulting equation the value of r^2 obtained from Eq. (445), we have

$$z = \frac{bw}{e(w^2 - e^2t^2)^{\frac{1}{2}}}. \quad \ldots \quad (446)$$

The equilibrium in this case evidently requires $w > et$.

If it be required to find the equation of the curve on all points of which m will be at rest, we have, from Eq. (441),

$$wz - tr = \text{a constant}, \quad \ldots \quad (447)$$

which may be written

$$wz - tr = (w - t)a'. \quad \ldots \quad (448)$$

By substituting for z its value $r \cos \theta$, this equation becomes

$$r = \frac{\left(1 - \dfrac{w}{t}\right)a'}{1 - \dfrac{w}{t} \cos \theta}, \quad \ldots \quad (449)$$

the polar equation of an ellipse, parabola, or hyperbola, according as w is less than, equal to, or greater than t, the pole being at the focus.

The Potential.

159. The general theory of attraction embraces the consideration of those forces by which matter attracts or repels other matter, and whose intensities are functions of the masses, and of the distances which separate them. The general term *attraction* is used for both repellent and attractive forces; the former being affected by the positive and the latter by the negative sign, to distinguish them.

In the following discussion the theory of attraction is limited to those forces whose intensities are expressed by the law, Eq. (2),

$$G = \frac{mm'}{r^2}\mu.$$

Electrical and magnetic forces are sometimes attractive and sometimes repellent; and since their intensities vary according to the law of the inverse square of the distance, Eq. (2) may also be applied to them by considering m and m' to be *quantities of free electricity* or *magnetism* instead of masses; the unit quantity being that which will attract an opposite or repel a similar unit quantity with a unit intensity, at a unit's distance apart. In the following discussion the term *mass* will then for convenience be taken to apply to the quantities m and m'. As in gravitation, these mutual attractions or repulsions are equal in intensity, opposite in direction, and exert their own influence whether other forces act on the masses or not.

160. *Component Attractions.*—Let m be the attracted mass, m' one of the attracting masses, r the distance between them, and x, y, z the co-ordinates of m' referred to m. We have then

$$r^2 = x^2 + y^2 + z^2; \quad \ldots \ldots \quad (450)$$

$$\frac{dr}{dx} = \frac{x}{r}, \quad \frac{dr}{dy} = \frac{y}{r}, \quad \frac{dr}{dz} = \frac{z}{r}; \quad \ldots \quad (451)$$

$$d\left(\frac{1}{r}\right) = -\frac{1}{r^2}dr. \quad \ldots \ldots \quad (452)$$

THE POTENTIAL.

The component attraction of m' for m in the direction of the co-ordinate axis x will then be

$$X = \frac{mm'}{r^2}\mu\frac{x}{r} = -mm'\mu\frac{d\left(\frac{1}{r}\right)}{dr}\frac{dr}{dx} = -mm'\mu\frac{d\left(\frac{1}{r}\right)}{dx}; \quad (453)$$

and similarly, for the other axes,

$$Y = -mm'\mu\frac{d\left(\frac{1}{r}\right)}{dy}; \quad Z = -mm'\mu\frac{d\left(\frac{1}{r}\right)}{dz}. \quad \ldots \quad (454)$$

The sums of the component attractions of all the molecules m', m'', m''', etc., for m are therefore

$$\left.\begin{aligned}X &= -m\mu\left(m'\cdot\frac{d\left(\frac{1}{r'}\right)}{dx} + m''\frac{d\left(\frac{1}{r''}\right)}{dx} + \text{etc.}\right)\\ &= -m\mu\frac{d\Sigma\frac{m'}{r}}{dx};\\ Y &= -m\mu\frac{d\Sigma\frac{m'}{r}}{dy};\\ Z &= -m\mu\frac{d\Sigma\frac{m'}{r}}{dz};\end{aligned}\right\} \cdot (455)$$

which, when we place

$$\Sigma\frac{m'}{r} = V, \quad \ldots \ldots \quad (456)$$

become

$$X = -m\mu\frac{dV}{dx};$$
$$Y = -m\mu\frac{dV}{dy};$$
$$Z = -m\mu\frac{dV}{dz}.$$
. . . . . (457)

That is, *the sum of the component attractions of the masses m', m'', etc., on m, in any direction, is equal to the corresponding partial differential coefficient of the function V multiplied by the product $m\mu$.*

161. The Potential.—The function V is called the *Potential* of the mass $\Sigma m'$ with reference to the *position* of m, and is defined to be the sum of the quotients obtained by dividing the mass or corresponding quantity of each element m' by its distance from m.

To explain what is meant by the potential, let $d\pi$ be the change in the potential energy of a *unit mass* attracted by m', under the assumed law of attraction, when the unit mass has changed its distance by dr; then we have

$$d\pi = -\frac{m'\mu}{r^2}dr; \quad \ldots \ldots (458)$$

whence

$$\pi = m'\mu\int -\frac{dr}{r^2} = \frac{m'\mu}{r} + C, \quad \ldots \ldots (459)$$

or, between the limits r and ∞,

$$\pi = m'\mu\int_\infty^r -\frac{dr}{r^2} = \frac{m'\mu}{r}.. \quad \ldots \ldots (460)$$

If the *unit mass* be subjected to the attraction of all the masses m', m'', etc., then we have

$$\Sigma \pi = \mu \Sigma \frac{m'}{r} = \mu V, \quad \ldots \ldots \quad (461)$$

which becomes, when the absolute intensity μ is taken to be unity,

$$\Sigma \pi = \Sigma \frac{m'}{r} = V. \quad \ldots \ldots \quad (462)$$

That is, *the potential is equal to the change in the potential energy of a unit mass when the latter is moved to infinite distance from any distance r against the decreasing attraction* $- \Sigma \frac{m'}{r^2}$, *or when the unit mass is brought from infinite distance to any distance r in opposition to the increasing repulsion,* $\Sigma \frac{m'}{r^2}$; *the absolute intensity being taken as unity in both cases.*

For any definite system of masses and forces governed by the assumed law *the potential at a point* has a definite value and can be expressed in units of work. When the potential is known the component attraction in any direction can be readily obtained by means of Eqs. (457).

162. *Equi-potential Surfaces.*—Let R be the intensity of the resultant attraction of the system of masses for m at the distance r; then, after multiplying Eqs. (457) by dx, dy, dz, respectively, and adding, we have

$$Xdx + Ydy + Zdz = - m\mu \left(\frac{dV}{dx}dx + \frac{dV}{dy}dy + \frac{dV}{dz}dz \right), \quad (463)$$

or

$$R dr = - m\mu dV, \quad \ldots \ldots \quad (464)$$

whence

$$R = -m\mu\frac{dV}{dr}. \quad \ldots \ldots \quad (465)$$

Let s be the path of m as it changes its position in any direction; θ the angle which s makes with r; and I the component of R in the direction of s. Then we have

$$I = R\cos\theta = -m\mu\frac{dV}{dr}\cos\theta = -m\mu\frac{dV}{ds}. \quad . \quad (466)$$

Hence *the component attraction in any direction varies directly with the first differential coefficient of the potential regarded as a function of the path in that direction.*

An *equi-potential* surface is one for which the potential is constant at all of its points. If the path s be a line of such a surface, we have

$$dV = -\frac{Ids}{m\mu} = -\frac{Rdr}{m\mu} = 0, \quad \ldots \ldots \quad (467)$$

and hence

$$V = -\frac{1}{m\mu}\int Ids = -\frac{1}{m\mu}\int Rdr = C. \quad . \quad . \quad (468)$$

A surface which fulfils for each of its points this condition is an equi-potential surface for the system of attractions. As any value may be attributed to C between its greatest and least values, there will be an indefinitely great number of equi-potential surfaces, corresponding to any given system of attractions, each of which will be a closed surface.

From Eq. (466) it is evident that I becomes zero when θ is 90°, and is the resultant attraction, or R, when θ is zero; hence the equi-potential surface cuts at right angles the direction of R at every point of its surface. These action lines of R are called *lines of force*, and any collection of them passing through an elementary portion of the surface is called a *tube of force*.

If the surface be supposed perfectly smooth, m would remain at rest on every point of it if subjected only to the system of attractions, and the surface would be pressed only in the direction of the normal. For this reason it is called a *level surface* or a *surface of equilibrium*. No two surfaces belonging to the same system can intersect or have a common point, for Eq. (468) cannot be satisfied for the same values of x, y, z, and give C dissimilar values. Of any two surfaces, the interior one corresponds to the greater resultant attraction and the greater value of the potential when the attraction is negative, and to the greater repulsion and greater value of the potential when the attraction is positive.

163. The determination of the values of the potential and the attractions for any given system of masses or quantities $\Sigma m'$ depends on the solution of the equations

$$V = \Sigma \frac{m'}{r}, \quad R = -m\mu \frac{dV}{dr} \quad \text{and} \quad I = -m\mu \frac{dV}{ds}. \quad (469)$$

When the quantities m' are elements of a quantity whose density and boundary vary by continuity, we may write

$$m' = dM' = \delta dv; \quad \ldots \ldots (470)$$

in which dv represents the elementary volume of M'. In such cases we have, when rectangular co-ordinates are employed,

$$V = \int \frac{\delta dv}{r} = \int \frac{\delta dx dy dz}{r}. \quad \ldots \ldots (471)$$

If it be desirable to use polar co-ordinates, we have for the value of the elementary volume, Fig. 59,

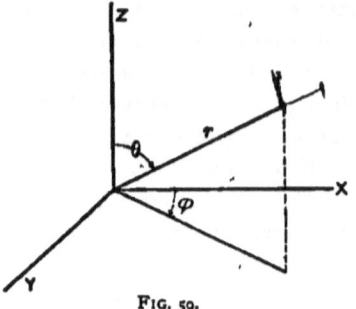

Fig. 59.

$$dv = dr \cdot r \sin\theta d\phi \cdot rd\theta \atop = r^2 dr \sin\theta d\phi d\theta; \qquad (472)$$

for, the edge of the infinitesimal cube in the direction of r is dr, the horizontal edge perpendicular to r is $r \sin\theta d\phi$, and that perpendicular to the plane of these two is $rd\theta$; and hence, in polar co-ordinates,

$$V = \int \delta r dr \sin\theta d\phi d\theta. \quad \ldots \quad (473)$$

164. Examples.—In the following examples the mass m will be taken to be *the unit mass*, and *the absolute intensity* μ to be *unity* also. By multiplying the results obtained by $m\mu$ we readily get the attractions for any mass and any intensity.

1. *The Potential and Attractions of a Straight Rod at an External Point.*—Let ω be the area of the rod's cross-section, y the distance of the external point m, Fig. 60, from the rod, and r its distance from any element of the rod; and let the axis of the rod be taken as the axis x, and the axis y pass through the external point. The element volume of the rod is then ωdx, and the distance $r = \sqrt{y^2 + x^2}$; then we have

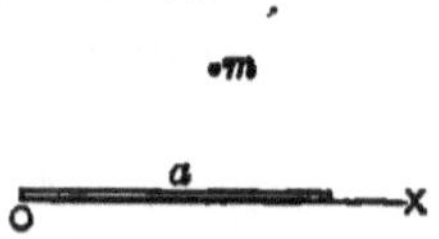

Fig. 60.

$$V = \delta\omega \int \frac{dx}{\sqrt{y^2 + x^2}} = \delta\omega \log \frac{x + \sqrt{y^2 + x^2}}{y}; \quad . \quad (474)$$

which, taken between the limits x'' and $-x'$, corresponding to the extremities of the rod, becomes

$$V = \delta\omega\left[\log\frac{x'' + \sqrt{y^2 + x''^2}}{y} - \log\frac{-x' + \sqrt{y^2 + x'^2}}{y}\right]$$

$$= \delta\omega \log\frac{x'' + \sqrt{y^2 + x''^2}}{-x' + \sqrt{y^2 + x'^2}}. \quad \ldots \quad \ldots \quad (475)$$

The component attraction in the direction of the rod is

$$X = \frac{dV}{dx} = \delta\omega\left(\frac{1}{\sqrt{y^2 + x''^2}} - \frac{1}{\sqrt{y^2 + x'^2}}\right). \quad \ldots \quad (476)$$

To find that perpendicular to the rod we have, from Eq. (474),

$$\frac{dV}{dy} = \delta\omega\frac{d\left[\log\left(\frac{x}{y} + \sqrt{1 + \frac{x^2}{y^2}}\right)\right]}{dy}$$

$$= -\delta\omega\frac{\dfrac{x}{y^2} + \dfrac{yx^2}{y^4\sqrt{1 + \dfrac{x^2}{y^2}}}}{\dfrac{x}{y} + \sqrt{1 + \dfrac{x^2}{y^2}}}$$

$$= -\delta\omega\frac{x}{y\sqrt{y^2 + x^2}} \quad \ldots \quad \ldots \quad (477)$$

Therefore between the limits x'' and $-x'$ we have

$$Y = \frac{dV}{dy} = \frac{\delta\omega}{y}\left(\frac{x''}{\sqrt{y^2 + x''^2}} + \frac{x'}{\sqrt{y^2 + x'^2}}\right). \quad \ldots \quad (478)$$

When the point m is on the perpendicular bisecting the rod, we have

$$X = 0; \quad Y = \frac{2\delta\omega}{y} \cdot \frac{x''}{\sqrt{y^2 + x''^2}} = \frac{2\delta\omega}{y} \sin \theta; \quad . \quad (479)$$

in which θ is the angle included between the bisecting perpendicular and the line drawn from m to the extremity of the rod.

2. *The Potential and Attractions of a Circular Arc at its Centre.*

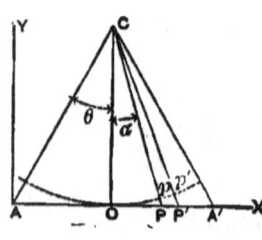

Fig. 61.

—Let θ be the angle subtended by any portion of the arc estimated from its middle point, Fig. 61. The element volume is then $\omega r d\theta$, and we have

$$V = \delta\omega \int_{-\theta'}^{\theta'} d\theta = 2\delta\omega\theta', \quad . \quad (480)$$

which is independent of the radius of the arc. The resultant attraction is evidently in the direction CO, or along the radius drawn to its middle point. Its value is

$$R = \delta\omega \int_{-\theta'}^{\theta'} \frac{r d\theta}{r^2} \cos \theta = \frac{2\delta\omega}{r} \sin \theta'. \quad . \quad . \quad (481)$$

From this we see that the attraction of the arc at the centre is the same as the attraction of the straight rod AOA'. Since also the masses of the elements pp' and PP' have the ratio

$$Cp : CP \sec \alpha = Cp : CP\frac{CP}{CO} = \overline{Cp}^2 : \overline{CP}^2, \quad . \quad . \quad (482)$$

their attractions on m at C are equal; whence, any portion of the right line tangent to the arc at O, as PP', attracts m at C with the same intensity as the corresponding arc pp'.

3. *The Potential and Attraction of a uniform Circular Ring at a Point on the Perpendicular to its Plane through its Centre.*—Let a be the radius of the ring, and c the distance of the point O, Fig. 62. Then

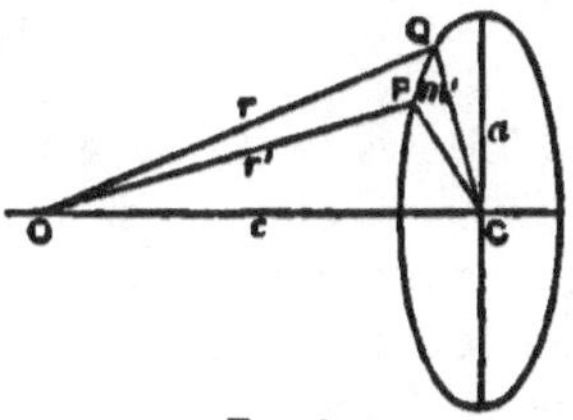

$$V = \Sigma \frac{m'}{r} = \frac{2\pi a \delta \omega}{\sqrt{a^2 + c^2}}. \quad . \quad (483)$$

When $c = 0$, this becomes a maximum, or $2\pi \delta \omega$, which, being independent of the radius, shows that V is a constant for all concentric rings at the centre.

Fig. 62.

As the sum of the component attractions of the elements of the ring in the plane of the ring is zero, the resultant attraction is in the direction of the perpendicular to the plane of the ring. Its value is evidently

$$R = \Sigma \frac{m'}{r^2} \cos QOC = \frac{c}{r^3} \Sigma m' = \frac{2\pi a c \delta \omega}{(a^2 + c^2)^{\frac{3}{2}}}. \quad . \quad (484)$$

4. *The Potential and Attraction of a Circular Plate at a Point on the Perpendicular to its Plane through its Centre.*—Let t be the thickness of the plate, Fig. 62, and suppose the plate made up of separate rings whose width is da. Then we have $\omega = tda$, and the element volume is $2\pi t a da$; whence

$$V = 2\pi \delta t \int_0^a \frac{a da}{\sqrt{a^2 + c^2}} = 2\pi \delta t (\sqrt{a^2 + c^2} - c); \quad . \quad (485)$$

which for the centre of the plate becomes

$$V = 2\pi \delta t a. \quad . \quad . \quad . \quad . \quad . \quad . \quad . \quad . \quad (486)$$

The resultant attraction is

$$R = \delta \int_0^a \frac{2\pi t a da}{r^2} \cos QOC = 2\pi \delta t c \int_0^a \frac{a da}{(a^2 + c^2)^{\frac{3}{2}}}$$
$$= 2\pi t c \delta \left(-\frac{1}{\sqrt{a^2 + c^2}}\right)_0^a = 2\pi \delta t \left(1 - \frac{c}{\sqrt{a^2 + c^2}}\right). \quad . \quad (487)$$

But since

$$\frac{c}{\sqrt{a^2 + c^2}} = \cos QOC,$$

it follows that the attraction of all circular plates of the same thickness and density, for a given molecule on a perpendicular to its plane through its centre, is the same for equal angles, subtended by the plate at the molecule. Therefore, if a molecule be at the vertex of a right cone with a circular base, the attractions of the normal sections of equal thickness of the cone for the molecule are equal. If the radius of the plate be infinite the attraction becomes $2\pi\delta t$, which is independent of the position of m with respect to the plate.

5. *The Potential and Attractions of a Spherical Shell at any Point.*—Taking the centre of the shell as the origin of a system of polar co-ordinates, let a be the radius and ρ the distance of the point from the centre; then the volume element is, Eq. (472),

$$a^2 t \sin \theta d\theta d\phi,$$

and

$$r = (a^2 - 2a\rho \cos \theta + \rho^2)^{\frac{1}{2}}; \quad \ldots \quad (488)$$

whence

$$V = \delta t a^2 \int_0^{2\pi} \int_0^{\pi} \frac{\sin \theta d\theta d\phi}{(a^2 - 2a\rho \cos \theta + \rho^2)^{\frac{1}{2}}} \quad \cdot \cdot \quad (489)$$

Integrating first with respect to ϕ, we have

$$V = 2\pi \delta t a^2 \int_0^{\pi} \frac{\sin \theta d\theta}{(a^2 - 2a\rho \cos \theta + \rho^2)^{\frac{1}{2}}}; \quad \cdot \cdot \quad (490)$$

and then with respect to θ,

$$V = \frac{2\pi \delta t a}{\rho} \left\{ (a^2 - 2a\rho \cos \theta + \rho^2)^{\frac{1}{2}} \right\}_0^{\pi}$$

$$= \frac{2\pi \delta t a}{\rho} [(a^2 + 2a\rho + \rho^2)^{\frac{1}{2}} - (a^2 - 2a\rho + \rho^2)^{\frac{1}{2}}]. \quad (491)$$

There are two cases to consider: (1) when m is within the surface, or $\rho < a$; and (2) when m is without the surface, or $\rho > a$. In the first case we have

$$V = \frac{2\pi\delta ta}{\rho}[a + \rho - (a - \rho)] = 4\pi\delta ta; \quad . \quad . \quad (492)$$

$$R = \frac{dV}{d\rho} = 0; \quad . \quad . \quad . \quad . \quad . \quad . \quad . \quad . \quad . \quad . \quad (493)$$

and in the second,

$$V = \frac{2\pi\delta ta}{\rho}[a + \rho - (\rho - a)] = \frac{4\pi\delta ta^2}{\rho} = \frac{M}{\rho}; \quad (494)$$

$$\frac{dV}{d\rho} = R = -\frac{4\pi\delta ta^2}{\rho^2} = -\frac{M}{\rho^2}; \quad . \quad . \quad . \quad . \quad . \quad (495)$$

M being the mass of the shell. Hence the potential of the interior space is constant and the resultant attraction zero, while all external space is made up of concentric spherical equi-potential surfaces, and the attraction at any point is the same in intensity and direction as though the whole quantity M were concentrated at the centre of the shell.

6. *The Potential and Attraction of a Thick Homogeneous Spherical Shell at any Point.*—Let the radii of the exterior and interior surfaces of the shell be a' and a'' respectively. The potential will, in general, consist of two parts, one corresponding to the shell within the spherical surface containing the point and the other to the shell without it.

For the first part we have (Ex. 5)

$$dV = \frac{4\pi\delta a^2}{\rho}da; \quad . \quad . \quad . \quad . \quad . \quad . \quad (496)$$

and for the second,

$$dV = 4\pi\delta a\, da; \quad . \quad . \quad . \quad . \quad . \quad (497)$$

hence

$$V = \frac{4\pi\delta}{\rho} \int_{a''}^{\rho} a^2 da + 4\pi\delta \int_{\rho}^{a'} a\, da$$
$$= \frac{4\pi\delta}{3\rho}(\rho^3 - a''^3) + 2\pi\delta(a'^2 - \rho^2). \quad . \quad . \quad (498)$$

If the point be wholly without the shell, then, M being the mass of the shell, we have

$$V = \frac{4\pi\delta}{\rho} \int_{a''}^{a'} a^2 da = \frac{4\pi\delta}{3\rho}(a'^3 - a''^3) = \frac{M}{\rho}; \quad (499)$$

$$R = \frac{M}{\rho^2}; \quad . \quad . \quad . \quad . \quad . \quad . \quad . \quad . \quad . \quad . \quad . \quad (500)$$

and if wholly within,

$$V = 2\pi\delta(a'^2 - a''^2); \quad . \quad . \quad . \quad . \quad . \quad . \quad . \quad (501)$$

$$R = 0. \quad . \quad . \quad . \quad . \quad . \quad . \quad . \quad . \quad . \quad . \quad . \quad (502)$$

If the attracting quantity be a homogeneous sphere $a'' = 0$, and at an interior point

$$V = 2\pi\delta a'^2 - \frac{2\pi\delta\rho^2}{3}; \quad . \quad . \quad . \quad . \quad . \quad . \quad . \quad (503)$$

$$R = \frac{4}{3}\pi\delta\rho = \frac{M}{a'^3}\rho; \quad . \quad . \quad . \quad . \quad . \quad . \quad . \quad (504)$$

and at an exterior point

$$V = \frac{4\pi\delta a'^3}{3\rho} = \frac{M}{\rho}; \quad . \quad . \quad . \quad . \quad . \quad . \quad . \quad (506)$$

$$R = \frac{M}{\rho^2}. \quad . \quad . \quad . \quad . \quad . \quad . \quad . \quad . \quad . \quad . \quad (507)$$

THE POTENTIAL.

For $\rho = a'$ we have, from both sets,

$$V = \frac{4\pi\delta a'^{2}}{3}; \quad \ldots \quad (508) \qquad R = \frac{M}{a'^{2}}; \quad \ldots \quad (509)$$

hence both V and R are continuous functions.

From this we see that, considering the earth as homogeneous, we may take its potential and attraction at any external point as though the whole mass of the earth were concentrated at its centre; while the attraction at an interior point is directly proportional to its distance from the centre.

165. *The Theorem of Laplace.*—Let S be any closed surface, and M any attracting quantity wholly external to S; let V be the potential of M, and ρ the normal to the surface reckoned outward. Then it is to be proved that

$$\int \frac{dV}{d\rho} dS = 0. \quad \ldots \quad \ldots \quad (510)$$

From any molecule m', Fig. 63, of M draw a right line piercing the surface S. It will pierce the surface in an even number

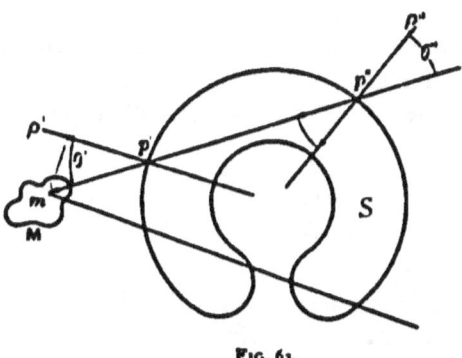

FIG. 63.

of points. Let p' and p'' be a pair of these points, θ' and θ'' the angles made by the normals at p' and p'' with the intersecting line, and r' and r'' the distances of p' and p'' from m'. Then we have

$$\frac{dV}{d\rho'} = \frac{m'}{r'^2}\cos\theta'; \quad \frac{dV}{d\rho''} = -\frac{m'}{r''^2}\cos\theta''. \quad \ldots \quad (511)$$

With m' as a vertex, conceive an indefinitely small cone whose solid angle is ω to be described about $m'p'$. The areas intercepted by the cone on S at the points p' and p'' are

$$dS' = \frac{\omega r'^2}{\cos\theta'} \quad \text{and} \quad dS'' = \frac{\omega r''^2}{\cos\theta''}; \quad \ldots \quad (512)$$

and hence

$$\frac{dV}{d\rho'}dS' = m'\omega; \quad \frac{dV}{d\rho''}dS'' = -m'\omega; \quad \ldots \quad (513)$$

$$\frac{dV}{d\rho'}dS' + \frac{dV}{d\rho''}dS'' = 0; \quad \ldots \quad (514)$$

and this result is true for every pair of points.

Now suppose the cone whose vertex is m' envelops the whole surface S; then its solid angle is made up of an indefinitely great number of elementary cones whose solid angles are ω, to each of which the above reasoning will apply. Therefore we have

$$\int \frac{dV}{d\rho}dS = 0. \quad \ldots \ldots \quad (515)$$

166. *Poisson's Extension of Laplace's Theorem.*—If any portion of M, as M', be contained within the closed surface S, then it is to be proved that

$$\int \frac{dV}{d\rho}dS = -4\pi M'. \quad \ldots \ldots \quad (516)$$

Let m' be the mass of one of the molecules of M', and let a right line be drawn from it, piercing S. It will pierce S in an odd number of points, which may be arranged in pairs as above, with one point remaining.

For the pairs Eq. (515) will be true, and for the odd point we have

$$\frac{dV}{d\rho}dS = -m'\omega. \quad \ldots \quad \ldots \quad (517)$$

Integrating for the whole space about m', we have

$$\int\frac{dV}{d\rho}dS = -m'\int_0^{4\pi}d\omega = -4\pi m'; \quad \ldots \quad (518)$$

and for all the molecules of M',

$$\int\frac{dV}{d\rho}dS = -4\pi M'. \quad \ldots \quad \ldots \quad (519)$$

Hence we conclude, Eqs. (515) and (519), that the sum of the attractions of a mass M, estimated outwardly at all points of a closed surface, is zero when the attracting matter M is wholly external to S, and is $-4\pi M'$ when the closed surface S contains any portion, M', of M.

167. This theorem is also expressed by the equations

$$\frac{d^2V}{dx^2} + \frac{d^2V}{dy^2} + \frac{d^2V}{dz^2} = 0 \quad \text{or} \quad = -4\pi\delta. \quad \ldots \quad (520)$$

To show this, let x, y, z be the co-ordinates of the attracting molecule, of density δ, without or within the surface, which may be taken as the surface of a rectangular parallelopipedon $dx\,dy\,dz$. Then

$$\int\frac{dV}{d\rho}dS,$$

for the face $dydz$, is

$$-\frac{dV}{dx}dydz;$$

and for the opposite face is

$$\left\{\frac{d^2V}{dx^2}dx + \frac{dV}{dx}\right\}dydz;$$

and for this pair of faces is

$$\frac{d^2V}{dx^2}dxdydz;$$

and similarly for the other faces we have

$$\frac{d^2V}{dy^2}dxdydz$$

and

$$\frac{d^2V}{dz^2}dxdydz.$$

Hence the integral

$$\int\frac{dV}{d\rho}dS = \left\{\frac{d^2V}{dx^2} + \frac{d^2V}{dy^2} + \frac{d^2V}{dz^2}\right\}dxdydz.$$

Placing the second member equal to 0 and $-4\pi M'$ in succession, we have, when the attracting matter is external to a closed surface,

$$\frac{d^2V}{dx^2} + \frac{d^2V}{dy^2} + \frac{d^2V}{dz^2} = 0; \quad \ldots \quad (521)$$

and when it contains it, wholly or in part,

$$\frac{d^2V}{dx^2} + \frac{d^2V}{dy^2} + \frac{d^2V}{dz^2} = -4\pi\delta. \quad \ldots \quad (522)$$

These theorems find their most frequent application in electricity and magnetism.

Motion of a System of Bodies.

168. The conclusions of Arts. 78 and 82 with respect to the motion of a single body under the action of extraneous forces are similarly true of a group or system of bodies when the motion of its centre of mass determines the translation of the system in space, and the rotation is estimated about that centre. To show this, let $x_{\prime}, y_{\prime}, z_{\prime}$, be the co-ordinates of the centre of mass of the system referred to any fixed origin; x, y, z, the co-ordinates of the centre of mass of each body referred to the fixed origin, and x', y', z', when referred to the centre of mass of the system; and let M be the type-symbol of the masses of the bodies. Applying Eqs. (T$_m$) to each mass and summing the results, we have

$$\left. \begin{array}{l} \Sigma X = \Sigma M \dfrac{d^2x}{dt^2}; \\[4pt] \Sigma Y = \Sigma M \dfrac{d^2y}{dt^2}; \\[4pt] \Sigma Z = \Sigma M \dfrac{d^2z}{dt^2}; \end{array} \right\} \quad \ldots \ldots \quad (523)$$

and since

$$x = x_{\prime} + x', \qquad y = y_{\prime} + y', \qquad z = z_{\prime} + z'. \quad (524)$$

$$d^2x = d^2x_{\prime} + d^2x', \quad d^2y = d^2y_{\prime} + d^2y', \quad d^2z = d^2z_{\prime} + d^2z', \quad (525)$$

these become, by applying the principle of the centre of mass,

$$\left.\begin{array}{l}\Sigma X = \dfrac{d^2 x_i}{dt^2}\Sigma M; \\ \Sigma Y = \dfrac{d^2 y_i}{dt^2}\Sigma M; \\ \Sigma Z = \dfrac{d^2 z_i}{dt^2}\Sigma M.\end{array}\right\} \quad \ldots \ldots \quad (526)$$

Similarly applying Eqs. (T_m') when impulsions alone act, we have

$$\left.\begin{array}{l}\dfrac{dx_i}{dt}\Sigma M = \Sigma M V_x; \\ \dfrac{dy_i}{dt}\Sigma M = \Sigma M V_y; \\ \dfrac{dz_i}{dt}\Sigma M = \Sigma M V_z;\end{array}\right\} \quad \ldots \ldots \quad (527)$$

hence the conclusions of Art. 78 follow from Eqs. (526) and (527) with respect to the centre of mass of the system.

169. If each of Eqs. (526) be multiplied in succession by the two co-ordinates which it does not contain, and the difference of the products be taken, then the summation of these differences for all the bodies of the system gives

$$\left.\begin{array}{l}\Sigma(Y x_i - X y_i) = \Sigma M \left(x_i \dfrac{d^2 y_i}{dt^2} - y_i \dfrac{d^2 x_i}{dt^2}\right); \\ \Sigma(X z_i - Z x_i) = \Sigma M \left(z_i \dfrac{d^2 x_i}{dt^2} - x_i \dfrac{d^2 z_i}{dt^2}\right); \\ \Sigma(Z y_i - Y z_i) = \Sigma M \left(y_i \dfrac{d^2 z_i}{dt^2} - z_i \dfrac{d^2 y_i}{dt^2}\right);\end{array}\right\} \quad \ldots \quad (528)$$

from which the motion of the centre of mass about the fixed origin may be found.

In the same way we have, from Eqs. (T_m),

$$\left. \begin{aligned} \Sigma(Yx - Xy) &= \Sigma M\left(x\frac{d^2y}{dt^2} - y\frac{d^2x}{dt^2}\right); \\ \Sigma(Xz - Zx) &= \Sigma M\left(z\frac{d^2x}{dt^2} - x\frac{d^2z}{dt^2}\right); \\ \Sigma(Zy - Yz) &= \Sigma M\left(y\frac{d^2z}{dt^2} - z\frac{d^2y}{dt^2}\right). \end{aligned} \right\} \quad \ldots \quad (529)$$

Substituting in these Eqs. (529) the values of the co-ordinates and accelerations from Eqs. (524) and (525) and reducing by the principle of the centre of mass, we obtain

$$\left. \begin{aligned} \Sigma(Yx' - Xy') &= \Sigma M\left(x'\frac{d^2y'}{dt^2} - y'\frac{d^2x'}{dt^2}\right); \\ \Sigma(Xz' - Zx') &= \Sigma M\left(z'\frac{d^2x'}{dt^2} - x'\frac{d^2z'}{dt^2}\right); \\ \Sigma(Zy' - Yz') &= \Sigma M\left(y'\frac{d^2z'}{dt^2} - z'\frac{d^2y'}{dt^2}\right). \end{aligned} \right\} \quad \ldots \quad (530)$$

Since these equations are independent of the co-ordinates of the centre of mass of the system, we conclude that the motion of rotation of the centres of mass of the constituent bodies about the centre of mass of the system is the same as if this point were at rest, and their motion is therefore entirely independent of the motion of translation of the latter point, a conclusion precisely similar to that of Art. 82.

170. *Conservation of the Motion of the Centre of Mass.*—If we suppose that a material system has been put in motion and then subjected only to the mutual attractions of its own bodies, we shall have

$$\Sigma X = 0, \quad \Sigma Y = 0, \quad \Sigma Z = 0; \quad \ldots \quad (531)$$

then there can be no accelerations of the centre of mass of the system, and Eqs. (526) become

$$\frac{d^2 x_i}{dt^2} = 0, \quad \frac{d^2 y_i}{dt^2} = 0, \quad \frac{d^2 z_i}{dt^2} = 0; \quad \ldots \quad (532)$$

from which we have

$$\frac{dx_i}{dt} = a, \quad \frac{dy_i}{dt} = b, \quad \frac{dz_i}{dt} = c; \quad \ldots \quad (533)$$

$$x_i = at + a', \quad y_i = bt + b', \quad z_i = ct + c'; \quad . \quad (534)$$

$$\frac{x_i - a'}{a} = \frac{y_i - b'}{b} = \frac{z_i - c'}{c}. \quad \ldots \quad (535)$$

That is, *if a system of masses be subjected only to its mutual attractions, its centre of mass will either be at rest or move uniformly in a right line.* This is called *the principle of the conservation of the motion of the centre of mass.*

171. If the masses of the solar system be subjected only to their mutual attractions of gravitation, the conditions of Eqs. (531) are satisfied for this system, and therefore its centre of mass must have uniform and rectilinear motion, or be at rest. Since the mass of the sun is very much greater than the sum of all the other masses of the constituents of the system, the error of assuming the centre of mass of the solar system to be coincident with that of the sun is slight. Calculations founded on the observations of astronomers show that this latter point is moving through space with a velocity of very nearly five miles per second, but sufficient data are not yet available to determine whether its path is a right line or an arc of small curvature; the latter being the more probable, owing to the extraneous forces of attraction of other systems.

By the same principle, the motion of the centre of mass of

the earth is uninfluenced by earthquakes or volcanic explosions occurring upon it, and that of the centre of mass of a projectile is not affected by its explosion, since the impulsive forces in each of these cases are mutually counterbalanced.

172. *Conservation of Moments, Invariable Axis, and Invariable Plane.* — If the forces acting on the system be the mutual attractions of its masses, we have the conditions

$$\left. \begin{array}{l} \Sigma(Yx' - Xy') = 0; \\ \Sigma(Xz' - Zx') = 0; \\ \Sigma(Zy' - Yz') = 0; \end{array} \right\} \quad \ldots \ldots \quad (536)$$

which reduce, Eqs. (530), to

$$\left. \begin{array}{l} \Sigma M\left(x'\dfrac{d^2y'}{dt^2} - y'\dfrac{d^2x'}{dt^2}\right) = 0; \\ \Sigma M\left(z'\dfrac{d^2x'}{dt^2} - x'\dfrac{d^2z'}{dt^2}\right) = 0; \\ \Sigma M\left(y'\dfrac{d^2z'}{dt^2} - z'\dfrac{d^2y'}{dt^2}\right) = 0; \end{array} \right\} \quad \ldots \ldots \quad (537)$$

and which, by integration, become

$$\left. \begin{array}{l} \Sigma M \dfrac{x'dy' - y'dx'}{dt} = C'; \\ \Sigma M \dfrac{z'dx' - x'dz'}{dt} = C''; \\ \Sigma M \dfrac{y'dz' - z'dy'}{dt} = C'''. \end{array} \right\} \quad \ldots \ldots \quad (538)$$

That is, when the forces acting on the system are the mutual attractions of its masses, *the algebraic sums of the moments of the moments of the masses of the system with respect to any set of rectangular co-ordinate axes at the centre of mass of the system are constant;* this is called *the principle of the conservation of moments.*

This principle may also be stated as follows: If the bodies of the system be supposed at rest in any one of its configurations, a definite system of impulsions would give each body its actual velocity. Eqs. (538) show that the sum of the component moments of these impulsions with respect to each co-ordinate axis is constant. The resultant moment of the system is also constant, and is given by

$$C = \sqrt{C'^2 + C''^2 + C'''^2}. \quad \ldots \quad (539)$$

As the co-ordinate planes may be assumed at pleasure, it is evident that the constants C', C'', C''' will in general change with each set of co-ordinate axes. The resultant axis of the system on which C would be measured is normal to that plane with reference to which the sum of the products of the projected areas by the masses is the maximum constant, and is the common intersection of those planes on which these sums are zero. This axis, and the normal plane through the centre of mass of the system, are called the *invariable axis* and *invariable plane* of the system of masses; the equation of the latter,

$$C'z' + C''y' + C'''x' = 0, \quad \ldots \quad (540)$$

is found by multiplying each of Eqs. (538) by the co-ordinate which it does not contain and adding the results together.

173. *Conservation of Areas.*—Eqs. (538) express another principle, which is known as *the conservation of areas*. Let radii-vectores r be drawn from the centre of mass of each body to that of the system, supposed at rest; then changing $x'dy' - y'dx'$ into its equivalent expression in polar co-ordinates with the pole at the centre of mass of the system, we have

$$\left. \begin{array}{ll} x' = r' \cos \theta, & dx' = dr' \cos \theta - r' \sin \theta d\theta; \\ y' = r' \sin \theta, & dy' = dr' \sin \theta + r' \cos \theta d\theta; \end{array} \right\} \quad \cdot \quad (541)$$

$$\therefore \quad x'dy' - y'dx' = r'^2 d\theta. \quad \ldots \quad \ldots \quad (542)$$

But $r'd\theta$ is twice the projection on the plane $x'y'$ of the sectoral area described by the radius vector of M' in the time dt; and the corresponding factors in the other equations are similarly the projections of double the differential areas on the planes $x'z'$ and $y'z'$ respectively. Let the type symbols of twice these projections on the planes $x'y'$, $x'z'$, $y'z'$ be denoted respectively by dA_z, dA_y, and dA_x; Eqs. (538) then become

$$\left. \begin{aligned} \Sigma M \frac{dA_z}{dt} &= C'; \\ \Sigma M \frac{dA_y}{dt} &= C''; \\ \Sigma M \frac{dA_x}{dt} &= C'''. \end{aligned} \right\} \quad \cdots \cdots (543)$$

Integrating between the limits corresponding to the interval t, we have

$$\left. \begin{aligned} \Sigma M A_z &= C't; \\ \Sigma M A_y &= C''t; \\ \Sigma M A_x &= C'''t. \end{aligned} \right\} \quad \cdots \cdots (544)$$

That is, *if a system of masses be subjected only to its mutual attractions, the sum of the products of each mass by the projection of its sectoral area about the centre of the system on any plane varies directly with the time.* This statement of the principle is called *the conservation of areas.*

If the resultant of a system of extraneous forces act through the centre of the system, Eqs. (536) will be satisfied and the conclusions of Arts. 170, 172 will apply to this case also.

174. *Relative Acceleration.*—If one of two bodies be supposed fixed and all the motion be attributed to the other, the acceleration which the latter would have under this supposition is called its *relative acceleration.* To find the relative acceleration of one body of the system with reference to the centre of mass of any other, let

M be the mass to which the motion is referred, and call this body the *central*; M', the mass of the moving body, called the *primary*; M'', the type-symbol of the masses of the remaining bodies of the system, called the *perturbating* bodies. Let the symbol (MM') represent the intensity of the reciprocal attraction of the masses M and M' along the line joining their centres, and the same symbol with a subscript letter, as $(MM')_x$, the component intensity in the direction of the corresponding axis; and similarly for the other masses and directions. Let x, y, z be the co-ordinates of M' referred to M. The component relative acceleration of M' with respect to M is the sum of their actual component accelerations due to their mutual attraction, plus the difference between the components of their actual accelerations due to the attractions of the perturbating bodies. The actual accelerations of M and M' due to their mutual attraction are $\dfrac{(MM')}{M}$ and $\dfrac{(MM')}{M'}$, and those due to one of the perturbating bodies are $\dfrac{(MM'')}{M}$ and $\dfrac{(M'M'')}{M'}$. The component relative accelerations are therefore

$$\left. \begin{aligned} \frac{d^2x}{dt^2} &= \left[\frac{(MM')_x}{M}+\frac{(MM')_x}{M'}\right]+\left[\frac{\Sigma(MM'')_x}{M}-\frac{\Sigma(M'M'')_x}{M'}\right]; \\ \frac{d^2y}{dt^2} &= \left[\frac{(MM')_y}{M}+\frac{(MM')_y}{M'}\right]+\left[\frac{\Sigma(MM'')_y}{M}-\frac{\Sigma(M'M'')_y}{M'}\right]; \\ \frac{d^2z}{dt^2} &= \left[\frac{(MM')_z}{M}+\frac{(MM')_z}{M'}\right]+\left[\frac{\Sigma(MM'')_z}{M}-\frac{\Sigma(M'M'')_z}{M'}\right]. \end{aligned} \right\} \quad (545)$$

175. The path of the centre of mass of the primary with respect to the central is called the *relative orbit* of the primary; its relative path influenced by the action of the perturbating bodies is called the *disturbed* or actual orbit, and if the action of these latter bodies be neglected the resulting relative path is called the *undisturbed* orbit. The differential equations in this case become

$$\left.\begin{aligned} \frac{d^2x}{dt^2} &= \mp \left[\frac{(MM')}{M} + \frac{(MM')}{M'}\right]\frac{x}{r}; \\ \frac{d^2y}{dt^2} &= \mp \left[\frac{(MM')}{M} + \frac{(MM')}{M'}\right]\frac{y}{r}; \\ \frac{d^2z}{dt^2} &= \mp \left[\frac{(MM')}{M} + \frac{(MM')}{M'}\right]\frac{z}{r}; \end{aligned}\right\} \quad \ldots \quad (546)$$

in which the upper sign corresponds to attraction and the lower to repulsion, and r is the radius vector of M'. When the law of the reciprocal attraction or repulsion is known the value of its intensity may be substituted for its symbol (MM'), and the resulting equations being integrated twice, there will result the component relative velocities and co-ordinates of the centre of mass of the primary referred to the centre of mass of the central body.

Central Forces.

176. A *central force* is one whose action-line is directed to or from a fixed point called the *centre of force*, and whose intensity is a function of the distance of the body acted on from that point. The force is attractive or repulsive according as its action-line is directed toward or from the centre.

177. *Laws of Central Forces.*—(1) Let the two masses M and M' be subjected to the action of their mutual attraction or repulsion. Then the motion of one, relative to the centre of mass of the other, may be considered as resulting from the action of a central force whose centre is the centre of mass of that body which is considered as fixed; Eqs. (546) are then applicable. Multiply the first by y and the second by x, and take the difference of the products; then multiply the first by z and the third by x, taking the difference of the products; and lastly, multiply the second by z and the third by y, taking the difference of the products, and we shall obtain

$$\left.\begin{array}{l} x\dfrac{d^2y}{dt^2} - y\dfrac{d^2x}{dt^2} = 0; \\ z\dfrac{d^2x}{dt^2} - x\dfrac{d^2z}{dt^2} = 0; \\ y\dfrac{d^2z}{dt^2} - z\dfrac{d^2y}{dt^2} = 0. \end{array}\right\} \quad \ldots \ldots (547)$$

Integrating, we have

$$\left.\begin{array}{l} x\dfrac{dy}{dt} - y\dfrac{dx}{dt} = h'; \\ z\dfrac{dx}{dt} - x\dfrac{dz}{dt} = h''; \\ y\dfrac{dz}{dt} - z\dfrac{dy}{dt} = h'''. \end{array}\right\} \quad \ldots \ldots (548)$$

Multiplying these equations by z, y and x, respectively, we have, by addition,

$$h'z + h''y + h'''x = 0, \quad \ldots \ldots (549)$$

the equation of an invariable plane. Hence *the orbit of a body acted on by a central force is contained in a fixed plane through the centre of force.*

(2) Take xy to be the plane of the orbit; then Eqs. (548) reduce to the single equation

$$x\dfrac{dy}{dt} - y\dfrac{dx}{dt} = h; \quad \ldots \ldots (550)$$

or, in polar co-ordinates,

$$r^2\dfrac{d\theta}{dt} = h; \quad \ldots \ldots (551)$$

in which r is the radius vector of M', θ is the variable angle made by r with a fixed line of reference, and h is double the sectoral area described by the radius vector in its own plane in the unit time.

Integrating Eq. (551) between the limits corresponding to values of the radius vector r' and r'' and the angles θ' and θ'', we have

$$\int_{r'\theta'}^{r''\theta''} r^2 d\theta = h(t' - t'') = ht; \quad \ldots \quad (552)$$

or, *the sectoral area described by the radius vector in the plane of the orbit varies directly with the time.*

Reciprocally, when this law is fulfilled we have, by differentiating Eq. (550) and multiplying by M',

$$M'\frac{d^2y}{dt^2}x - M'\frac{d^2x}{dt^2}y = Yx - Xy = 0; \quad \ldots \quad (553)$$

and the orbit is described under the action of a central force.

(3) From Eq. (551) we have

$$\frac{d\theta}{dt} = \frac{h}{r^2}; \quad \ldots \ldots \ldots \quad (554)$$

or, *the angular velocity of the body in its orbital motion about the centre of force varies inversely as the square of the radius vector.*

(4) Since the velocity of the body in its orbit is

$$v = \frac{ds}{dt} = \frac{ds}{d\theta}\frac{d\theta}{dt}, \quad \ldots \ldots \quad (555)$$

we have, from Eq. (554),

$$v = \frac{h}{r^2}\frac{ds}{d\theta} \cdot \quad \ldots \ldots \ldots \quad (556)$$

Let p be the length of the perpendicular from the centre of force to the tangent to the orbit at the body's place; then

$$p = r\sin\phi = r\frac{rd\theta}{ds}, \qquad (557)$$

and hence

$$v = \frac{h}{p}; \qquad (558)$$

or, *the velocity of the body varies inversely as the perpendicular distance from the centre of force to the tangent to the orbit at the place of the body.*

(5) Let A be the relative acceleration at any point of the orbit, and ρ the corresponding radius of curvature; then the component relative acceleration in the direction of ρ is

$$A\frac{p}{r} = \frac{V^2}{\rho}; \qquad (559)$$

from which we have

$$V^2 = 2A \times \frac{1}{4}\left(2\rho\frac{p}{r}\right).. \qquad (560)$$

But $2\rho\frac{p}{r}$ is the length of the chord of curvature drawn through the centre of force to the place of the body. Comparing Eq. (560) with $V^2 = 2gh$, we see that the *actual velocity of the body at any point of its orbit is that due to a height equal to one fourth of the chord of curvature drawn through the centre of force; the body starting from rest and the intensity of the central force remaining constant over this distance.* If the orbit be circular, R its radius, and its centre coincide with the centre of force, the velocity becomes constant and

$$V^2 = AR. \qquad (561)$$

The acceleration in the direction of the tangent to the orbit is

$$\frac{d^2s}{dt^2} = -A\frac{dr}{ds}. \quad \ldots \ldots \quad (562)$$

Multiplying by M' and ds and integrating between limits, we have

$$\tfrac{1}{2}M'(V_1^2 - V_2^2) = -M'\int_{r_2}^{r_1} A\,dr, \quad \ldots \quad (563)$$

the equation of energy. Hence the orbital velocity is independent of the path described and varies with the distance of the body from the centre of force. In any closed orbit, therefore, when the body returns successively to the same position in its orbit, the velocity will always be the same as before.

These are the general laws of central forces, and are seen to be independent of the character and law of variation of the central force.

178. *The Differential Equation of the Orbit.*—Assuming the co-ordinate plane xy as the plane of the orbit and employing polar co-ordinates, we have, from the law of areas, Eq. (551),

$$r^2\frac{d\theta}{dt} = k,$$

which becomes, when r is replaced by $\frac{1}{u}$,

$$\frac{d\theta}{dt} = ku^2. \quad \ldots \ldots \quad (564)$$

Differentiating the following equation:

$$x = r\cos\theta = \frac{\cos\theta}{u}; \quad \ldots \ldots \quad (565)$$

and dividing by dt, we obtain

$$\frac{dx}{dt} = -\frac{u\sin\theta + \cos\theta\frac{du}{d\theta}}{u^2}\cdot\frac{d\theta}{dt}$$

$$= -h\left(u\sin\theta + \cos\theta\frac{du}{d\theta}\right), \quad \ldots \quad (566)$$

and hence

$$\frac{d^2x}{dt^2} = -h\left(u\cos\theta + \cos\theta\frac{d^2u}{d\theta^2}\right)\frac{d\theta}{dt}$$

$$= -h^2u^2\left(u\cos\theta + \cos\theta\frac{d^2u}{d\theta^2}\right). \quad \ldots \quad (567)$$

Placing this value equal to that of the relative acceleration $\frac{d^2x}{dt^2}$ in the first of Eqs. (546), we have, after dividing by $\cos\theta$ and remembering that $\frac{x}{r} = \cos\theta$,

$$\frac{(MM')}{M'} + \frac{(MM')}{M} = A = \mp h^2u^2\left(u + \frac{d^2u}{d\theta^2}\right), \quad \ldots \quad (568)$$

the differential polar equation of the orbit of a body under the action of a central force.

179. To solve the direct and inverse problems in the case of a central force we proceed as follows:

(1) To find *the equation of the orbit*, substitute for A in Eq. (568) its value in terms of u, and integrate twice; the resulting equation expressing the relation between u and θ is the equation of the orbit. The two arbitrary constants which appear in the integration are determined by the initial conditions, viz., the initial values of the radius vector and velocity, and the initial direction of motion.

(2) To find *the law of the force* necessary to cause a body to

describe a given orbit, differentiate the polar equation of the orbit twice and substitute the resulting value of $\frac{d^2u}{d\theta^2}$ in Eq. (568); then eliminate θ, and the result will give A in terms of r.

180. *Particular Cases of the Direct Problem.*—(1) *To find the orbit due to an attractive central force whose intensity varies directly with the distance of the body from the centre.* Let the centre of force be the origin; μ', the measure of the intensity of the central attraction for a unit mass at a unit's distance. From the law of the force and the differential equation of the orbit we have

$$A = \mu' r = \frac{\mu'}{u} = h^2 u^2 \left(u + \frac{d^2 u}{d\theta^2} \right), \quad \ldots \quad (569)$$

or

$$\frac{d^2 u}{d\theta^2} + u = \frac{\mu'}{h^2 u^3}. \quad \ldots \quad (570)$$

Multiplying by $2du$ and integrating, we get

$$\frac{du^2}{d\theta^2} + u^2 = -\frac{\mu'}{h^2 u^2} + C. \quad \ldots \quad (571)$$

Let R be the initial value of the radius vector, and take the initial direction of motion perpendicular to R. Then at the time $t = 0$ we have $\frac{du}{d\theta} = 0$ and $u = \frac{1}{R}$; hence $C = \frac{1}{R^2} + \frac{\mu' R^2}{h^2}$. and

$$\frac{du^2}{d\theta^2} + u^2 = \frac{1}{R^2} + \frac{\mu' R^2}{h^2} - \frac{\mu'}{h^2 u^2}. \quad \ldots \quad (572)$$

Let V be the initial value of the velocity, and from Eq. (558) we have $h = RV$. Substituting this value of h in Eq. (572),

14

multiplying through by u^2, and changing the form of the resulting equation, we have

$$u^2\frac{du^2}{d\theta^2} + \left(u^2 - \frac{V^2 + \mu'R^2}{2V^2R^2}\right)^2 = \left(\frac{V^2 - \mu'R^2}{2V^2R^2}\right)^2. \quad \ldots \quad (573)$$

Solving with reference to $d\theta$, and multiplying by 2, we have

$$2d\theta = \frac{2udu}{\sqrt{\left(\frac{V^2 - \mu'R^2}{2V^2R^2}\right)^2 - \left(u^2 - \frac{V^2 + \mu'R^2}{2V^2R^2}\right)^2}}$$

$$= \frac{\frac{2V^2R^2}{V^2 - \mu'R^2}2udu}{\sqrt{1 - \left(\frac{2V^2R^2u^2 - (V^2 + \mu'R^2)}{V^2 - \mu'R^2}\right)^2}} \quad \ldots \quad (574)$$

Integrating between limits corresponding to $t = 0$ and t, remembering that the initial radius vector coincides with R, we have

$$2\theta = \sin^{-1}\frac{2V^2R^2u^2 - (V^2 + \mu'R^2)}{V^2 - \mu'R^2} - \sin^{-1}1; \quad (575)$$

whence we have

$$\sin(90° + 2\theta) = \cos 2\theta = \frac{2V^2R^2u^2 - (V^2 + \mu'R^2)}{V^2 - \mu'R^2}. \quad (576)$$

Clearing of fractions and changing to rectangular co-ordinates, we have, recalling that

$$\cos^2\theta = \frac{x^2}{x^2 + y^2} \quad \text{and} \quad \sin^2\theta = \frac{y^2}{x^2 + y^2},$$

$$2V^2R^2 - (V^2 + \mu'R^2)(x^2 + y^2) = (V^2 - \mu'R^2)(x^2 - y^2); \quad (577)$$

whence we have

$$\frac{x^2}{R^2} + \frac{\mu' y^2}{V^2} = 1, \qquad (578)$$

the equation of an ellipse referred to its centre and axes; hence *the orbit of a body acted on by a central attraction varying directly with the distance, is an ellipse whose centre is at the centre of force.*

To find the velocity at any point of the orbit. Let v be the general value of the velocity of the body in its orbit; then from Eq. (558) we have

$$v^2 = \frac{h^2}{p^2}. \qquad (579)$$

and from Eq. (556) we get

$$\frac{1}{p^2} = \frac{ds^2}{r^4 d\theta^2} = \frac{r^2 d\theta^2 + dr^2}{r^4 d\theta^2} = u^2 + \frac{du^2}{d\theta^2}. \qquad (580)$$

Hence

$$v^2 = h^2\left(u^2 + \frac{du^2}{d\theta^2}\right). \qquad (581)$$

But from Eq. (572) we have

$$h^2\left(u^2 + \frac{du^2}{d\theta^2}\right) = V^2 + \mu' R^2 - \frac{\mu'}{u^2}, \qquad (582)$$

and therefore

$$v^2 = V^2 + \mu'(R^2 - r^2), \qquad (583)$$

which gives v when r is known.

To find the Periodic Time. The semi-axes of the orbit are, Eq. (578), R and $\dfrac{V}{\sqrt{\mu'}}$. The periodic time, or the time required for the body to complete its orbit, is, by the law of areas, therefore

$$T = \frac{\pi R \frac{V}{\sqrt{\mu'}}}{\frac{h}{2}} = \frac{2\pi R V}{R V \sqrt{\mu'}} = \frac{2\pi}{\sqrt{\mu'}}. \quad \ldots \quad (584)$$

Hence the periodic time is independent of the dimensions of the orbit.

Examples of orbits under this law are found in molecular vibrations; in the small vibrations of elastic bodies, such as tuning-forks, stretched strings, etc.; and in the oscillations of a pendulum through small arcs.

181. (2) If the central force be repellent, the discussion above may be made applicable by changing the sign of μ', Eqs. (578); the equation of the orbit then becomes

$$\frac{x^2}{R^2} - \frac{\mu' y^2}{V^2} = 1; \quad \ldots \quad \ldots \quad (585)$$

hence *the orbit is an hyperbola whose centre is at the centre of force.*

182. (3) *To find the orbit when the central force is attractive and varies inversely as the square of the distance.* Assume the same notation as in the previous problem, and let the direction of the initial velocity make any angle with the prime radius vector R.

Then we have

$$A = \frac{\mu'}{r^2} = \mu' u^2 = h^2 u^2 \left(\frac{d^2 u}{d\theta^2} + u \right), \quad \ldots \quad (586)$$

or

$$\frac{d^2 u}{d\theta^2} + u = \frac{\mu'}{h^2}. \quad \ldots \quad \ldots \quad (587)$$

Multiplying by $2du$ and integrating, we have

$$\frac{du^2}{d\theta^2} + u^2 = \frac{2\mu' u}{h^2} + C. \quad \ldots \quad (588)$$

From the initial conditions and Eq. (581) we have

$$C = \frac{V^2}{h^2} - \frac{2\mu'}{h^2 R}; \quad \ldots \quad (589)$$

and therefore

$$\frac{du^2}{d\theta^2} + u^2 = \frac{V^2}{h^2} + \frac{2\mu' u}{h^2} - \frac{2\mu'}{h^2 R}. \quad \ldots \quad (590)$$

This equation may be written

$$\frac{du^2}{d\theta^2} = c^2 - (u - b)^2, \quad \ldots \quad (591)$$

by assuming

$$\frac{\mu'}{h^2} = b \quad \text{and} \quad \frac{V^2 R - 2\mu'}{R h^2} + \frac{\mu'^2}{h^4} = c^2. \quad \ldots \quad (592)$$

From Eq. (591) we have, taking the negative sign of the radical,

$$\frac{-du}{\sqrt{c^2 - (u - b)^2}} = d\theta; \quad \ldots \quad (593)$$

and by integration,

$$\cos^{-1}\frac{u - b}{c} = \theta + \gamma. \quad \ldots \quad (594)$$

in which γ, the constant of integration, is the initial angle which the prime radius vector makes with the fixed line of reference. From Eq. (594) we get

$$u = b + c \cos(\theta + \gamma), \quad \ldots \ldots (595)$$

or

$$r = \frac{1}{b + c \cos(\theta + \gamma)}; \quad \ldots \ldots (596)$$

and substituting the values of b and c,

$$r = \frac{\frac{h^2}{\mu'}}{1 + \sqrt{\frac{(V^2R - 2\mu')h^2}{R\mu'^2} + 1} \cdot \cos(\theta + \gamma)}, \quad (597)$$

which may be written

$$r = \frac{\frac{-R\mu'}{V^2R - 2\mu'}\left[1 - \left(\frac{(V^2R - 2\mu')h^2}{R\mu'^2} + 1\right)\right]}{1 + \sqrt{\frac{(V^2R - 2\mu')h^2}{R\mu'^2} + 1} \cdot \cos(\theta + \gamma)}. \quad (598)$$

Comparing this with the polar equation of a conic section referred to the focus as a pole,

$$r = \frac{a(1 - e^2)}{1 + e \cos \phi}, \quad \ldots \ldots (599)$$

we see that (598) is the equation of a conic section referred to the focus as a pole, in which

$$a = \frac{-R\mu'}{V^2R - 2\mu'}; \quad \ldots \ldots \ldots (600)$$

$$e = \sqrt{\frac{(V^2R - 2\mu')h^2}{R\mu'^2} + 1}; \quad \ldots \ldots (601)$$

$$\phi = \theta + \gamma. \quad \ldots \ldots \ldots \ldots (602)$$

CENTRAL FORCES.

Therefore the orbit will be an *ellipse, parabola* or *hyperbola* according as

$$V^2 R - 2\mu' < 0, \ = 0 \ \text{or} \ > 0;$$

that is, as

$$V < \sqrt{\frac{2\mu'}{R}}, \ = \sqrt{\frac{2\mu'}{R}} \ \text{or} \ > \sqrt{\frac{2\mu'}{R}}. \quad \ldots \quad (603)$$

183. To determine the meaning of $\sqrt{\frac{2\mu'}{R}}$, we have

$$\frac{d^2 r}{dt^2} = -\frac{\mu'}{r^2}. \quad \ldots \quad \ldots \quad (604)$$

Multiplying by $2dr$ and integrating between the limits $r = R$ and $r = \infty$, we have

$$\frac{dr^2}{dt^2} = v^2 = \frac{2\mu'}{R}, \quad \ldots \quad \ldots \quad (605)$$

or

$$v = \sqrt{\frac{2\mu'}{R}}. \quad \ldots \quad \ldots \quad (606)$$

Thus $\sqrt{\frac{2\mu'}{R}}$ is shown to be the velocity which the body would have if it should move from rest at infinity to the distance R under the action of the central force; it is called *the velocity from infinity at the distance R*.

Hence we conclude that *the orbit of a body under the action of a central attraction varying inversely as the square of the distance will be an ellipse, parabola or hyperbola according as the initial velocity is less than, equal to or greater than the velocity from infinity at the initial point.*

184. *To find the velocity at any point of the orbit.* For the velocity at any point we have, Eq. (581),

$$v^2 = h^2\left(\frac{du^2}{d\theta^2} + u^2\right), \quad \ldots \ldots (607)$$

and hence from Eq. (590) we have

$$v^2 = V^2 + 2\mu' u - \frac{2\mu'}{R} \quad \ldots \ldots (608)$$

or

$$v^2 = V^2 + 2\mu'\left(\frac{1}{r} - \frac{1}{R}\right), \quad \ldots \ldots (609)$$

from which the velocity corresponding to any radius vector r can be found.

We also see from Eq. (609) that the velocity at any point of the orbit will always conform to that which characterizes the particular orbit in question; that is, if the orbit be a parabola, for example, the velocity at any point whose radius vector is r will always be equal to $\sqrt{\frac{2\mu'}{r}}$ at that point, and, similarly, less than $\sqrt{\frac{2\mu'}{r}}$ for the ellipse and greater than $\sqrt{\frac{2\mu'}{r}}$ for the hyperbola.

185. *To find the time of description of any portion of the orbit.*

(1) *The Elliptical Orbit.*—We have for the equation of the orbit

$$\frac{1}{r} = u = \frac{1 + e \cos \theta}{a(1 - e^2)}; \quad \ldots \ldots (610)$$

hence

$$\frac{du}{d\theta} = -\frac{e \sin \theta}{a(1 - e^2)} \quad \ldots \ldots (611)$$

and
$$\frac{d^2u}{d\theta^2} = -\frac{e\cos\theta}{a(1-e^2)}. \quad \ldots \ldots (612)$$

Therefore
$$A = h^2u^2\left(\frac{d^2u}{d\theta^2} + u\right) = \frac{h^2u^2}{a(1-e^2)}; \quad \ldots (613)$$

and since
$$A = \frac{\mu'}{r^2},$$

we have
$$\mu' = \frac{h^2}{a(1-e^2)}, \quad \ldots \ldots (614)$$

or
$$h = \sqrt{\mu'a(1-e^2)}. \quad \ldots \ldots (615)$$

Therefore the periodic time is
$$T = \frac{2\pi a^2\sqrt{1-e^2}}{\sqrt{\mu'a(1-e^2)}} = 2\pi\sqrt{\frac{a^3}{\mu'}}. \quad \ldots (616)$$

We also have, from Eqs. (551) and (615),
$$dt = \frac{r^2}{h}d\theta = \frac{r^2 d\theta}{\sqrt{\mu'a(1-e^2)}}. \quad \ldots (617)$$

Differentiating Eq. (610), and substituting for sin θ its value deduced from the same equation, we have

$$d\theta = \frac{adr\sqrt{1-e^2}}{r\sqrt{a^2e^2 - (r-a)^2}}. \quad \ldots (618)$$

Substituting in Eq. (617), we have

$$dt = \sqrt{\frac{a}{\mu'}} \frac{r\,dr}{\sqrt{a^2e^2 - (r-a)^2}} = \sqrt{\frac{a}{\mu'}} \frac{[(r-a)+a]d(r-a)}{\sqrt{a^2e^2 - (r-a)^2}}. \quad (619)$$

Integrating between the limits corresponding to the nearer vertex $r_0 = a(1 - e)$ and any value of r, we have

$$t = \sqrt{\frac{a}{\mu'}} \left(a \sin^{-1}\frac{r-a}{ae} - \sqrt{a^2e^2 - (r-a)^2} \right)\Big|_{r_0}^{r}$$

$$= \sqrt{\frac{a}{\mu'}} \left(-a \sin^{-1}\frac{a-r}{ae} - \sqrt{a^2e^2 - (a-r)^2} \right)\Big|_{r_0}^{r}$$

$$= \sqrt{\frac{a}{\mu'}} \left(a \cos^{-1}\frac{a-r}{ae} - \sqrt{a^2e^2 - (a-r)^2} \right), \quad (620)$$

from which the time of description of any portion of the orbit can be found.

Making $r = a(1 + e)$, corresponding to the farther vertex, we have for the semi-periodic time

$$t = \pi\sqrt{\frac{a^3}{\mu'}}. \quad \cdots \quad (621)$$

Making $r = a$, corresponding to the extremity of the conjugate axis, we have

$$t = \left(\frac{\pi}{2} - e\right)\sqrt{\frac{a^3}{\mu'}}; \quad \cdots \quad (622)$$

and hence for the time from the extremity of the conjugate to the farther extremity of the transverse axis we have

$$t = \left(\frac{\pi}{2} + e\right)\sqrt{\frac{a^3}{\mu'}}. \quad \cdots \quad (623)$$

From these values we see that the velocity decreases from the nearer to the farther extremity of the transverse axis, and then increases to the nearer extremity.

The trajectory of a projectile in vacuo, under the supposition that its weight acts with constant intensity in parallel directions, was shown in Art. 93 to be a parabola. The preceding discussion shows that this trajectory will be an arc of an ellipse having the earth's centre at the farther focus, when gravity is considered as a central force varying inversely as the square of the distance from the centre of the earth. The transverse axis of the ellipse is the vertical through the highest point of the trajectory.

(2) *The Parabolic Orbit.*—We have for the equation of the orbit

$$r = \frac{2a}{1 + \cos \theta}, \quad \ldots \ldots (624)$$

in which $2a$ is the semi-parameter. We therefore have

$$A = h^2 u^2 \left(\frac{d^2 u}{d\theta^2} + u \right) = \frac{h^2 u^3}{2a} = \frac{h^2}{2a} \frac{1}{r^3}; \quad \ldots (625)$$

hence

$$\mu' = \frac{h^2}{2a}, \quad \text{or} \quad h^2 = 2a\mu', \quad \ldots \ldots (626)$$

and

$$v^2 = \frac{h^2}{p^2} = h^2 \left(u^2 + \frac{du^2}{d\theta^2} \right) = \frac{2\mu'}{r}. \quad \ldots \ldots (627)$$

We have from Eq. (551)

$$dt = \frac{r^2}{h} d\theta;$$

hence

$$t = \frac{1}{h}\int_{\theta_1}^{\theta_2} r^2 d\theta = \frac{1}{h}\int_{\theta_1}^{\theta_2} 4a^2\left(\frac{d\theta}{(1+\cos\theta)^2}\right) \quad \ldots \quad (628)$$

$$= \sqrt{\frac{a^3}{2\mu'}}\int_{\theta_1}^{\theta_2} \frac{4d\theta}{(1+\cos\theta)^2}$$

$$= \sqrt{\frac{a^3}{2\mu'}}\int_{\theta_1}^{\theta_2} \sec^4\frac{\theta}{2} d\theta$$

$$= \sqrt{\frac{2a^3}{\mu'}}\left(\tan\frac{\theta_2}{2} - \tan\frac{\theta_1}{2} + \frac{1}{3}\tan^3\frac{\theta_2}{2} - \frac{1}{3}\tan^3\frac{\theta_1}{2}\right). \quad (629)$$

Place $\tan\frac{\theta_2}{2} = t_2$ and $\tan\frac{\theta_1}{2} = t_1$, then the equation above becomes

$$t = \sqrt{\frac{2a^3}{\mu'}}\left(t_2 - t_1 + \frac{t_2^3 - t_1^3}{3}\right)$$

$$= \sqrt{\frac{2a^3}{\mu'}}(t_2 - t_1)\left(1 + \frac{t_2^2 + t_2 t_1 + t_1^2}{3}\right). \quad \ldots \quad (630)$$

Let $y^2 = 1 + \frac{(t_2 + t_1)^2}{4}$, and we have

$$t = \sqrt{\frac{2a^3}{\mu'}}(t_2 - t_1)\left(y^2 + \frac{(t_2 - t_1)^2}{12}\right)$$

$$= \frac{1}{3}\sqrt{\frac{2a^3}{\mu'}}\left[\left(y + \frac{t_2 - t_1}{2}\right)^3 - \left(y - \frac{t_2 - t_1}{2}\right)^3\right]. \quad \ldots \quad (631)$$

Let c be the length of the chord joining the extremities of the radii vectores r_2 and r_1; then we have

$$c^2 = r_2^2 - 2r_2 r_1 \cos(\theta_2 - \theta_1) + r_1^2$$
$$= (r_2 \cos\theta_2 - r_1 \cos\theta_1)^2 + (r_2 \sin\theta_2 - r_1 \sin\theta_1)^2. \quad (632)$$

CENTRAL FORCES.

We have also, since $r = \dfrac{2a}{1+\cos\theta} = \dfrac{a}{\cos^2 \tfrac{1}{2}\theta} = a(1+\tan^2 \tfrac{1}{2}\theta)$,

$$r_2 = a(1+t_2^2), \quad r_1 = a(1+t_1^2); \quad \ldots \quad (633)$$

$$\cos\theta_2 = \dfrac{1-t_2^2}{1+t_2^2}, \quad \sin\theta_2 = \dfrac{2t_2}{1+t_2^2}; \quad \ldots \quad (634)$$

$$\cos\theta_1 = \dfrac{1-t_1^2}{1+t_1^2}, \quad \sin\theta_1 = \dfrac{2t_1}{1+t_1^2}. \quad \ldots \quad (635)$$

Therefore

$$c^2 = 4a^2(t_2 - t_1)^2 \left(\dfrac{(t_2+t_1)^2}{4} + 1\right)$$
$$= 4a^2(t_2 - t_1)^2 y^2; \quad \ldots \ldots \ldots (636)$$

hence

$$c = 2a(t_2 - t_1)y, \quad \ldots \ldots \ldots (637)$$

and

$$r_2 + r_1 + c = 2a\left\{1 + \dfrac{t_2^2 + t_1^2}{2} + (t_2 - t_1)y\right\}$$
$$= 2a\left\{y + \dfrac{t_2 - t_1}{2}\right\}^2; \ldots \ldots (638)$$

and similarly

$$r_2 + r_1 - c = 2a\left\{y - \dfrac{t_2 - t_1}{2}\right\}^2 \ldots \ldots (639)$$

Substituting in Eq. (631), we have finally

$$t = \dfrac{1}{6\sqrt{\mu}}\left\{\sqrt{(r_2+r_1+c)^3} - \sqrt{(r_2+r_1-c)^3}\right\}; \quad (640)$$

from which t can be found in terms of the radii vectores and the chord of the parabolic arc.

(3) *The Hyperbolic Orbit.*—We have for the equation of the orbit

$$r = \frac{a(e^2 - 1)}{1 + e \cos \theta},$$

or

$$u = \frac{1}{a(e^2 - 1)} + \frac{e}{a(e^2 - 1)} \cos \theta; \quad \ldots \quad (641)$$

and therefore

$$t = \int_0^\theta \frac{r^2 d\theta}{\sqrt{\mu' a(e^2 - 1)}}$$

$$= \sqrt{\frac{a}{\mu'}} \int_{a(e-1)}^r \frac{r \, dr}{\sqrt{(r+a)^2 - a^2 e^2}}$$

$$= \sqrt{\frac{a}{\mu'}} \left\{ \sqrt{(r+a)^2 - a^2 e^2} - a \log \frac{r + a + \sqrt{(r+a)^2 - a^2 e^2}}{ae} \right\}; \quad (642)$$

from which the time corresponding to the description of that part of the orbit from the vertex to the point corresponding to any radius vector r can be found.

186. *The Anomalies.*—When the central attraction varies inversely as the square of the distance, the position of the body in its orbit is generally referred to the right line coinciding with the least radius vector. This line is called *the line of apsides*, and its intersections with the orbit are called *apses;* the one nearer to the focus being the *lower*, and the other the *higher*, *apsis*. The angle included between the line of apsides and the radius vector is called the *anomaly*, and is measured from the lower apsis as an origin.

Let us place

$$\cos^{-1} \frac{a - r}{ae} = u, \quad \ldots \quad \ldots \quad (643)$$

an auxiliary angle; then we have

$$r = a(1 - e \cos u). \quad \ldots \quad \ldots \quad (644)$$

Substituting the value of u in Eq. (620), we have, after placing $\sqrt{\dfrac{a^3}{\mu'}} = \dfrac{1}{n}$,

$$nt = u - e \sin u. \quad \ldots \quad (645)$$

Equating the values of r from Eqs. (644) and (610), we have

$$\frac{1 - e^2}{1 + e \cos \theta} = 1 - e \cos u, \quad \ldots \quad (646)$$

whence

$$\left.\begin{array}{l} 1 + \cos \theta = \dfrac{(1 - e)(1 + \cos u)}{1 - e \cos u}; \\[1em] 1 - \cos \theta = \dfrac{(1 + e)(1 - \cos u)}{1 - e \cos u}; \end{array}\right\} \quad \ldots \quad (647)$$

and therefore

$$\frac{1 - \cos \theta}{1 + \cos \theta} = \frac{1 + e}{1 - e} \cdot \frac{1 - \cos u}{1 + \cos u}, \quad \ldots \quad (648)$$

or

$$\tan \tfrac{1}{2}\theta = \sqrt{\frac{1 + e}{1 - e}} \tan \tfrac{1}{2} u. \quad \ldots \quad (649)$$

The angle θ is called the *true anomaly*, u the *eccentric anomaly*, and nt the *mean anomaly*.

From Eqs. (644), (645) and (649) the values of the true anomaly θ and the radius vector r can be found in terms of the eccentricity and the mean anomaly. (See Price's Calculus, vol. iii. pp. 561–567.) These are

$$\left.\begin{array}{l} \theta = nt + 2e \sin nt + \dfrac{5}{4} e^2 \sin 2nt + \dfrac{e^3}{12}(13 \sin 3nt - 3 \sin nt) + \text{etc.}; \\[1em] r = a\left(1 - e \cos nt + \dfrac{e^2}{2}(1 - \cos 2nt) - \dfrac{3e^3}{8}(\cos 3nt - \cos nt) + \text{etc.}\right). \end{array}\right\} (650)$$

Hence, knowing n the mean motion, the time since the epoch, and the eccentricity, the *true anomaly* or the angular distance of the body from the line of apsides, and the *distance of the body from the focus* at once results. The difference between the true and mean anomaly $\theta - nt$, called the *Equation of the Centre*, is evidently a function of the eccentricity, and its value is obtained at any time t from the first of Eqs. (650).

187. To illustrate the geometrical meaning of these quantities let $APB'A$, Fig. 64, be the elliptical orbit, S the centre of force at the focus, P the position of the body, $SP = r$, $PSA = \theta$ and $AC = a$. On AA' describe a semicircle ABA'. From the properties of the ellipse we have

$$SP = a - eCM, \qquad (651)$$

and therefore

$$r = a - ae \cos QCM$$
$$= a - ae \cos u. \qquad (652)$$

From Eq. (616) we see that the periodic time is independent of the eccentricity of the ellipse; it is therefore the same as that in the circle whose radius is a; but in this case $e = 0$, $r = a$, $\theta = u = nt$. Hence nt represents the arc of the circle which would be described uniformly by a body in the same time as that in which the elliptic arc is described, both bodies starting from A, and both reaching A' at the same time; n is therefore called the mean motion of the body. Since sin u is positive in the first two quadrants, we see, Eq. (645), that u is greater than nt while the body is describing that part of its orbit from A to A', and less than nt from A' to A; therefore the true place of the body is in advance of its mean place in the first and second quadrants, and behind in the third and fourth; nt is

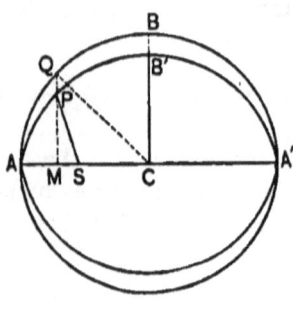

Fig. 64.

therefore called the mean anomaly, and since u depends on the value of e, u is called the eccentric anomaly. Both the velocity and the angular velocity are greatest at A and least at A', as is seen from the equations giving these values in the laws of central forces.

The Solar System.

188. *The Solar System* consists of the sun and other bodies whose relative positions and motions are mutually dependent, and which taken together may be considered as a single system of bodies in space. It derives its name from the sun, the great central body about which all the other members of the system, called *primary* or *secondary* bodies, revolve.

The *primary* bodies are:

(1) The four *inner or lesser planets*, Mercury, Venus, the Earth and Mars, named in order of their distance from the sun.

(2) A group of minor planets called *Asteroids*, of which over two hundred and sixty have so far been noted and catalogued.

(3) The four *outer or greater planets*, Jupiter, Saturn, Uranus, and Neptune.

(4) A number of *Comets* and *Meteors*, or bodies having masses much smaller, and generally orbits of much greater eccentricity, than those of the planets and asteroids.

The *secondary* bodies are the *Satellites* or *Moons* of the planets, which describe orbits about the latter and are carried with them in orbital motion about the sun; of these now known *three* belong to the inner and *seventeen* to the outer planets.

All the bodies of the solar system are spheroidal in form, and their diameters are very small compared with the distances which separate them from each other. In addition to their orbital motions they have a motion of rotation about their axes.

The mass of the sun is more than seven hundred and forty times as great as the sum of the masses of all the other bodies of the system. Owing to this fact, and to the relative positions of the planets, the centre of mass of the entire system lies within the sun's volume and not far from its own centre.

189. *Kepler's Laws*.—John Kepler, of Wurtemburg, was the first to announce the laws governing the motion of the planets about the sun. This announcement was the result of more than twenty years' faithful and laborious study of the observations collected by his predecessor Tycho Brahe. Kepler's investigations were principally directed to the explanation of the apparent irregularities of the motion of the planet Mars, whose orbit was at that time supposed to be an epicycloid. These laws of Kepler not only completely accounted for the motion of Mars, but also satisfactorily explained the motions of all the other planets about the sun, and those of the satellites about their respective primaries. These laws are:

(1) *The orbit of each planet about the sun is an ellipse, having one of its foci in the sun's centre.*

(2) *The areas described by the radius vector of each planet in its orbital motion vary directly as the times of describing them.*

(3) *The squares of the periodic times of the planets are directly proportional to the cubes of their mean distances from the sun's centre.*

190. If we assume, for the present, that these laws are accurately true, we readily deduce the following consequences, viz.:

(1) The orbit of each planet being an ellipse having the sun's centre in one focus, it follows that the value of the relative acceleration becomes, Eq. (613),

$$A = \frac{h^2 u^2}{a(1-e^2)} = \frac{h^2}{a(1-e^2)} \frac{1}{r^2}; \quad \ldots \quad (653)$$

that is, *the relative acceleration varies inversely as the square of the distance of the planet from the centre of the sun.*

(2) From the second law, or that of equal areas described in equal times, we have

$$r^2 d\theta = h dt = x dy - y dx. \quad \ldots \quad (654)$$

Whence, after differentiating and dividing by dt^2, we have

$$x\frac{d^2y}{dt^2} - y\frac{d^2x}{dt^2} = 0. \quad \ldots \ldots (655)$$

Multiplying by M', the mass of the planet, we have

$$M'\frac{d^2y}{dt^2}x - M'\frac{d^2x}{dt^2}y = Yx - Xy = 0. \quad \ldots (656)$$

Therefore *the action-line of the reciprocal attraction of the sun and the planet always passes through the sun's centre;* and since this reciprocal attraction varies inversely as the square of the distance, *the force which keeps the planet in its orbit is a central force.*

(3) From Eq. (616) we have for the square of the periodic time of one of the planets whose mean distance is a' and whose mass is M'

$$T'^2 = \frac{4\pi^2 a'^3}{\mu'}; \quad \ldots \ldots (657)$$

in which μ' is the intensity of the central attraction for a unit mass at a unit's distance. Similarly for another planet whose distance is a'' and mass M'' we have

$$T''^2 = \frac{4\pi^2 a''^3}{\mu''}; \quad \ldots \ldots (658)$$

whence we have

$$\frac{T'^2}{T''^2} = \frac{a'^3}{a''^3} \cdot \frac{\mu''}{\mu'}. \quad \ldots \ldots (659)$$

From Kepler's third law, we have

$$\frac{T'^2}{T''^2} = \frac{a'^3}{a''^3} \cdot \quad \ldots \ldots (660)$$

which being substituted in the preceding equation gives

$$\mu'' = \mu'.$$

That is, the rigid truth of Kepler's third law involves the equality of the central attraction for all the planets. But

$$\mu' = (M + M')\mu \quad \text{and} \quad \mu'' = (M + M'')\mu; \qquad (661)$$

hence, since the masses of the planets are known to be unequal, we must conclude that Kepler's third law is not rigidly true.

If the masses of the sun, 1,000,000,000, of Jupiter, 954,305, the greatest, and of Mercury, 200, the least of the planets, be substituted in (661), we find

$$\frac{\mu''}{\mu'} = 1.000954, \qquad (662)$$

and this ratio will be more nearly equal to unity for any other pair of planets. The discrepancy in assuming unity for the ratio $\frac{\mu''}{\mu'}$ for the planets of the solar system is therefore in general negligible, and the consequence of Kepler's third law may be taken as true within sensible limits.

191. *Law of Gravitation*.—Later and more accurate observations than those which Kepler employed show that his laws are not exactly true, but are only very close approximations to the truth. The single law which governs planetary motions and definitely fixes their actual departures from the positions assigned by Kepler's laws is that of universal gravitation, which is thus enunciated by Isaac Newton:

That every particle of matter in the universe attracts every other particle, with an intensity which varies directly as the product of their masses, and inversely as the square of the distance which separates them.

Newton deduced this law from his investigations of the relative acceleration of the moon, in a direction normal to its orbit about the earth. He proved that the earth's relative attraction on the moon caused it to fall towards the earth, with an acceleration due to gravity, modified only by the increased distance and greater mass of the moon, precisely as a body near the earth's surface falls with its particular acceleration; and therefore concluded that the law of attraction between the earth and moon was essentially the same as that between the earth and body. From this deduction the generalization to the enunciated law of gravitation followed.

The intensity of the reciprocal attraction between the sun and a planet, whose masses are M and M' respectively, under the law of gravitation is therefore

$$G = \frac{MM'}{r^2}\mu; \quad \ldots \ldots \quad (663)$$

μ in this expression being the intensity of the reciprocal attraction of a unit mass for another unit mass at a unit's distance. Therefore the partial differential equations of the undisturbed orbit of a planet about the sun, Eqs. (546), become, under the law of gravitation,

$$\left. \begin{array}{l} \dfrac{d^2x}{dt^2} = -\dfrac{M+M'}{r^3}\mu x; \\[4pt] \dfrac{d^2y}{dt^2} = -\dfrac{M+M'}{r^3}\mu y; \\[4pt] \dfrac{d^2z}{dt^2} = -\dfrac{M+M'}{r^3}\mu z. \end{array} \right\} \quad \ldots \ldots \quad (664)$$

Kepler's first and second laws can be deduced directly from these equations, and hence are simply the consequences derived from the *undisturbed orbit* of a primary about the sun under the supposition that the law of gravitation is the governing law of their mutual attraction.

The differential equations of the *actual* or *disturbed orbit* can be obtained immediately from Eqs. (545), by substituting the values which the symbols (MM'), (MM'') and $(M'M'')$ take under the law of gravitation, and the resulting equations will differ from (664) only in the third and fourth terms. The latter being computed and applied properly to the undisturbed orbit will give the actual orbit of the planet. These terms, called the *perturbating functions* of the orbit, depend upon the relative attractions of the other planets for the sun and for the planet whose orbit is to be determined. Owing to the relatively great mass of the sun, and to the immense distances which separate the planets from each other, the *perturbations* or actual displacements of a planet from its undisturbed orbit are sensibly infinitesimal quantities, compared with the actual distance of the planet from the sun. Hence the component rectangular displacements due to the perturbating action of each planet may be computed for each planet separately as if it alone acted; then the algebraic sum of the separate perturbations in any direction may be taken as the resultant perturbating effect in that direction due to the simultaneous action of all the planets, without the least appreciable error. Because of this fact, the problem is called the *problem of three bodies*, viz., the sun, the planet and the perturbating body.

The theoretical deductions which flow from the assumption of the law of universal gravitation as the governing law of planetary motion have been amply confirmed by the accurate astronomical predictions of the positions and motions of the planetary bodies, made years in advance, and markedly so, by the circumstances attending the discovery of the planet Neptune; so that the law itself is at present accepted as the fundamental law of physical astronomy.

192. *Planetary Orbits.*—The undisturbed orbit of each of the bodies of the solar system has been shown to be a plane curve, whose plane passes through the sun's centre; but it is ascertained, by observation, that no two of these planes are coincident. In order to find the relative positions and motions of the bodies of the solar system at any time, it is necessary to refer them to the surface of the celestial sphere by some system of spherical co-ordinates.

The celestial sphere is usually taken to be that sphere which is enclosed by the surface of the visible heavens; but it may be taken to be any sphere whose centre is the position of the observer, and whose radius is entirely arbitrary. The *Ecliptic* is the great circle of intersection of the celestial sphere, by the plane of the earth's orbit. The *Celestial Equator*, or *Equinoctial*, is the great circle of intersection of the celestial sphere, by the plane of the earth's equator. The poles of the heavens are the poles of the Equinoctial, or are the points in which the earth's axis produced pierces the celestial sphere. The *Equinoxes* are the points in which the equinoctial and ecliptic intersect; the *Vernal Equinox* being that point in which the sun appears at the beginning of spring, and the *Autumnal Equinox* that in which it appears in the beginning of autumn. Celestial *longitude* and *latitude* are spherical co-ordinates by which any point is referred to the plane of the Ecliptic, and to that of a great circle of the celestial sphere perpendicular to the ecliptic, passing through the vernal equinox. *Celestial longitude* is the angular distance from the vernal equinox, measured on the ecliptic eastwardly in direction, to the circle of latitude which passes through the position in question; and *celestial latitude* is the angular distance to the given point, from the ecliptic, measured on that circle of latitude which passes through the given point. The line of intersection of the plane of a planet's orbit with the plane of the ecliptic is called the *line of nodes;* the *ascending node* being the point of the planet's orbit at which the planet passes from south to north of the ecliptic, the other being the *descending node*. The nearest point of a planet's orbit to the sun is called *perihelion*, and the farthest is called *aphelion*.

The *elements of a planet's orbit* are seven in number, viz.:

(1) The *inclination* of its plane to the plane of the ecliptic.

(2) The *longitude* of the *ascending node*.

(3) The *orbit longitude* of *perihelion*.

(4) The *mean distance* of the planet from the sun, or the *semi-transverse axis* of the planet's orbit.

(5) The *eccentricity* of the orbit.

(6) The *position* of the planet at any given time, as at the *epoch*.
(7) The *mean orbital motion*.

The first two of these elements fix the plane of the orbit with reference to the plane of the ecliptic; the third fixes the position of the line of apsides in this plane and from which the anomalies are reckoned; from the fourth and fifth the form and dimensions of the ellipse are determined; and the last two, called the *elements of position*, locate the planet in its orbit at any time.

The elements of any planetary orbit are deduced from three consecutive observations of its position in right ascension and declination and the times of observation, by methods which will be explained in the course in Astronomy.

Since the mean motion depends on the periodic time, and the latter, by Eq. (616), depends on the mass of the planet, it is necessary to explain how the mass of a planet is ascertained. The masses of the planets are so small, in comparison with the mass of the sun, that their values cannot be ascertained from Eq. (657) after substituting the observed periodic times and mean distances. But if we consider the planet Jupiter and one of its satellites, we will have, for the periodic time T' of the latter about Jupiter,

$$T'^2 = \frac{4\pi^2 a'^3}{M' + m}; \quad \ldots \ldots \quad (665)$$

in which a' is its mean distance from Jupiter, and m its mass; μ, the attraction of a unit mass at a unit's distance, being here taken as unity. Similarly, for the periodic time of Jupiter about the sun, we have

$$T^2 = \frac{4\pi^2 a^3}{M + M'}; \quad \ldots \ldots \quad (666)$$

whence we have

$$\frac{T^2}{T'^2} = \frac{a^3}{a'^3} \frac{M' + m}{M + M'}$$

$$= \frac{a^3}{a'^3} \frac{M'}{M}; \quad \ldots \ldots \quad (667)$$

if m be supposed small compared with M', and M' small compared with M. Substituting the known values of T, T', a and a', we have the ratio of the mass of Jupiter to that of the sun. In the same way the masses of all the planets having satellites may be compared with that of the sun; and if the mass of one of these be found, that of the remaining planets will at once result. To find the mass of the planets *Mercury* and *Venus*, which have no satellites, recourse must be had to their perturbating effects on the other planets.

The mass of the earth has been ascertained by direct measurement of its figure, magnitude and density. From the direct geodesic measurement of the arcs of the meridian in England, France, Russia, India and Africa, the form and dimensions of the earth have been determined. The density has been directly investigated by means of Dr. Maskelyne's observations with the pendulum near Schehallien Mountain in Scotland; and also by the experiments of Cavendish and Bailey from the attraction of leaden balls on small masses. From these, the mass of the earth having been found, the masses of the sun and the other planets are readily obtained. The further discussion of planetary motions is reserved for the course in Astronomy.

THEORY OF MACHINES.

193. Thus far we have regarded bodies as rigid solids, either wholly or partially free to move under the action of extraneous forces. But the various devices or machines designed for the transfer of energy from one system of masses to another are made up of parts which are neither free nor rigid; and in their use certain resistances are developed by the active extraneous forces, which make the actual results differ from the theoretical more or less widely. It is the office of *Experiment* to find the values of these resistances, to tabulate the results, and to deduce therefrom the *experimental laws* covering all cases that may arise in practice. We will consider briefly the resistances of *Friction* and *Stiffness of Cordage*.

Friction.

194. *Friction* is the resistance which the surface of one body offers to the sliding or rolling upon it of any other body. It is due to the roughness of the surfaces of bodies; for as no degree of polish can make any surface perfectly smooth, there will always be minute projections on one surface which interlock with those of the other. These projections must be broken down, abraded or lifted over each other before motion can take place. When the roughness of any two surfaces is diminished by polish or lubricants the friction between them decreases.

The friction which opposes a change of the body from rest to motion is called *static* friction, and that which accompanies motion is called *kinetic* friction. The latter may be either sliding or rolling; thus a heavy body dragged over a surface, an axle turning in a journal-box, and a vertical shaft turning on a horizontal plane, give examples of sliding friction; while a wheel rolling over the surface of the ground is resisted by rolling friction. Sliding friction, being the more common, will be alone discussed.

The *action-line* of friction coincides with the tangent to the surfaces at the point of contact, and its *direction* is always opposite to that of the motion. The *intensity* of friction must always be determined by experiment, or may be assumed from previous experiments on similar bodies and surfaces. Such results have been generalized into what are known as the *laws of friction*. The accepted laws have been deduced from the experiments made by Coulomb in 1781, and from those made by Morin under the direction of the French Government in 1830-4. They are:

(1) *The intensity of friction, for the same material surfaces, varies directly with the normal pressure.*

(2) *The intensity of friction is independent of the area of contact of the surfaces.*

(3) *The intensity of friction is independent of the velocity of motion of the rubbing surfaces.*

Recent experiments indicate that the last law is only approximately true for velocities less and greater than those employed in the experiments of Morin; and since the laws are wholly experimental, we are warranted in accepting them only within the limits covered by the experiments from which they were deduced. In static friction there are also variations depending on the length of time during which the surfaces have been in contact, and therefore greater discrepancies occur in this kind than in kinetic friction.

195. *Coefficient of Friction.*—If N represent the normal pressure, F the intensity of friction for any two surfaces due to N, and f the intensity of friction for a unit of normal pressure, we have, from the first law,

$$F = fN, \quad \text{or} \quad f = \frac{F}{N} \quad . \quad . \quad . \quad . \quad (668)$$

f is called the *coefficient of friction*, and when known for any two surfaces the total friction for those surfaces can be found for any normal pressure. Its value depends upon the nature of the rubbing surfaces, upon their smoothness, and upon the degree of lubrication given to them.

To find f experimentally in any particular case, let the body be placed upon a plane surface, Fig. 65, and let the latter be gradually inclined to the horizon until the body is in a state bordering on motion; then if motion be given the body, it will descend the plane uniformly. The forces are the weight of the body acting vertically downward, and the friction resisting up the plane. The normal component of the weight, if α be the angle made by the plane with the horizon, is $W \cos \alpha$, and hence the friction is $fW \cos \alpha$. The component of the weight parallel to the plane urges the body down the plane, and is $W \sin \alpha$; and since the motion is uniform the intensity of the parallel component must be equal to that of the resistance due to friction. Hence we have

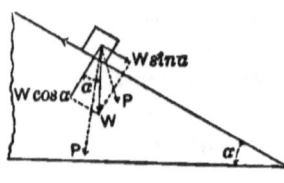

Fig. 65.

$$fW \cos \alpha = W \sin \alpha, \quad \ldots \ldots \quad (669)$$

and therefore

$$f = \tan \alpha. \quad \ldots \ldots \ldots \quad (670)$$

That is, *the coefficient of friction for any two materials is equal to the natural tangent of the inclination to the horizon which a plane of one of the substances must make in order that a body of the other substance may descend uniformly due to its weight when resisted only by friction.*

The inclination of the plane is usually called the *angle of friction*, and sometimes the *limiting angle of resistance*. To explain the meaning of this latter term, let a body rest on a plane inclined at the angle of friction; it is then said to be in a state bordering on motion downward. The angle at the centre of gravity of the body, made by the direction of the weight and the normal to the surface, is then equal to α, the angle of friction. If any force P *whose action-line lies within the angle* α be supposed introduced at the centre of gravity, it is evident that the additional developed friction arising from the normal component of P is greater than the component of P parallel to the plane; therefore the body will be no longer in a state bordering on motion. If the action-line of a force fall outside of the same angle on the side of the weight, the additional friction developed by its normal component will be less than the intensity of its parallel component; therefore the body will move down the plane with accelerated motion. If the action-line of any force coincide with that of the weight, the body will still be in a state bordering on motion downward, and if put in motion it will descend the plane uniformly.

196. *Problems involving Friction.*—(1) *Motion on a Plane Surface.* Let a body resting on a plane, Fig. 66, be acted on by its weight, W, and let any extraneous force whose action-line lies in a vertical plane, and whose intensity is I, be applied to it; let i be the inclination of the plane to the horizon, and θ the angle made by the force with the plane. If the body be bordering on motion downward the friction will act up the plane, and we shall have, from the principles of equilibrium,

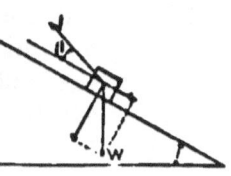

Fig. 66.

$$f(W\cos i - I\sin\theta) = W\sin i - I\cos\theta, \quad . \quad . \quad (671)$$

whence

$$I = W\frac{\sin i - f\cos i}{\cos\theta - f\sin\theta} \quad . \quad . \quad . \quad . \quad (672)$$

If the body be bordering on motion up the plane, the action-line of friction will change direction by 180°, and the value of I can be obtained from the preceding by changing the sign of f; or

$$I = W\frac{\sin i + f\cos i}{\cos\theta + f\sin\theta} \quad . \quad . \quad . \quad (673)$$

Therefore the value of I may vary between these limits without causing motion in the body.

Considering I a function of θ, we have, after differentiating the last equation,

$$\frac{dI}{d\theta} = W(\sin i + f \cos i)\frac{\sin \theta - f \cos \theta}{(\cos \theta + f \sin \theta)^2} \cdot \cdot \cdot (674)$$

Applying the condition for maxima and minima, by placing the second member equal to zero, we find

$$f = \tan \theta, \quad \text{or} \quad \theta = \tan^{-1} f, \quad \ldots \quad (675)$$

which corresponds to a minimum value of I; therefore a force is applied to the best advantage in moving a body up a plane when its action-line makes an angle with the plane equal to the angle of friction. As this result is independent of i, it is true whatever be the inclination of the plane to the horizon, and is therefore true if the plane be horizontal.

(2) *To find the Friction on a Trunnion.*—The cylindrical projections at the extremities of an axle are called *trunnions;* the cylindrical box upon which a trunnion is supported is called a *trunnion-bed* or *pillow-block*. When the axle is supported on its end, the latter is called a *pivot*.

Let A, Fig. 67, be the trunnion, B the trunnion-bed, and C any element of contact during rotation; let R be the resultant of all the extraneous forces acting on the trunnion, excluding friction; N and T the normal and tangential components of R respectively. If the trunnion rotate, it will rise in its box until the developed friction F is equal to T, after which sliding will occur at the element of contact. Then the resultant of R and F will be normal to the surface of the trunnion at the

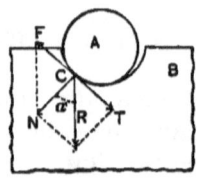

FIG. 67.

element of contact, and will be equal to N. Let α be the angle between R and N; then we have

$$N = R \cos \alpha \quad \text{and} \quad T = R \sin \alpha = F; \quad \ldots \quad (676)$$

whence

$$\frac{F}{N} = f = \tan \alpha. \quad \ldots \ldots \quad (677)$$

Therefore the element of contact during rotation is that at which the normal to the trunnion makes an angle with the resultant equal to the angle of friction.

To find the friction we have

$$\cos^2 \alpha = \frac{1}{1 + \tan^2 \alpha} = \frac{1}{1 + f^2}; \quad \ldots \quad (678)$$

multiplying by $f^2 R^2$ and extracting the square root, we have

$$F = fN = fR \cos \alpha = fR \frac{1}{\sqrt{1 + \tan^2 \alpha}} = R \frac{f}{\sqrt{1 + f^2}}. \quad (679)$$

That is, the friction on trunnions is equal to the resultant of the extraneous forces multiplied by $\dfrac{f}{\sqrt{1 + f^2}}$, in which f is the coefficient of friction for the materials which compose the trunnion and box.

The moment of friction on trunnions, if r be the radius of the trunnion, is

$$Rr \frac{f}{\sqrt{1 + f^2}}, \quad \ldots \ldots \quad (680)$$

and the work consumed by friction in n complete turns is

$$R 2\pi r n \frac{f}{\sqrt{1 + f^2}}. \quad \ldots \ldots \quad (681)$$

From the second law of friction we see that the intensity of friction and the work are independent of the length of the trunnion.

(3) *Friction on a Circular Pivot.*—Let the shaft be vertical and its pivot end a circle of radius a, Fig. 68, resting on a horizontal surface; and let the centre of the circle of contact be the origin of the polar co-ordinates r and θ, which fix the position of any elementary area of contact.

F<small>IG</small>. 68. Let p be the intensity of the normal pressure on each unit of area, which is equal to the whole normal pressure divided by the area of the pivot surface.

The expression for any elementary area is

$$rdrd\theta;$$

the normal pressure upon it is

$$prdrd\theta;$$

the developed friction is

$$fprdrd\theta;$$

and the moment of this friction with respect to the axis of the shaft is

$$fpr^2drd\theta.$$

Integrating this last expression between proper limits, we have for the resultant moment of the friction on a pivot

$$M = \int_0^{2\pi}\int_0^a fpr^2 dr d\theta = \tfrac{2}{3}fp\pi a^3. \quad \ldots \quad (682)$$

If N represent the whole normal pressure, we have

$$N = p\pi a^2, \quad \ldots \ldots (683)$$

whence

$$M = fN\tfrac{2}{3}a. \quad \ldots \ldots (684)$$

Therefore the moment of friction on a pivot is equal to the product of the coefficient of friction for the materials of which the shaft and support are made, the total normal pressure, and two thirds of the radius of the pivot surface.

From this we see that the moment of friction may be diminished by decreasing the circular area of the pivot, provided the diminished area be sufficient to withstand the normal pressure without penetrating the surface on which it rests. The distance $\tfrac{2}{3}a$ is called the *mean lever* of friction on pivots, and we see that if the whole friction be supposed concentrated on an arc of this radius its moment will be the same as the resultant moment of all the elementary frictions.

The work consumed by friction in n revolutions is

$$2\pi n f N \tfrac{2}{3}a. \qquad \qquad (685)$$

(4) *Friction on a Ring Pivot.*—Let the inner and outer radii of the ring be a and a' respectively; then (Eqs. (682) and (683))

$$M = \tfrac{2}{3} f p \pi (a'^3 - a^3), \qquad \qquad (686)$$

$$N = p\pi(a'^2 - a^2); \qquad \qquad (687)$$

whence

$$M = \tfrac{2}{3} f N \frac{a'^3 - a^3}{a'^2 - a^2}. \qquad \qquad (688)$$

If b be the breadth of the ring and r be its mean radius, we have

$$a' = r + \tfrac{1}{2}b,$$
$$a = r - \tfrac{1}{2}b;$$

which substituted in Eq. (688) give

$$M = fN\left(r + \frac{1}{12}\frac{b^2}{r}\right). \qquad \qquad (689)$$

The factor $\left(r + \dfrac{1}{12}\dfrac{b^2}{r}\right)$ is called the *mean lever* of friction for a ring pivot.

The quantity of work consumed by friction on pivots in n revolutions of the shaft, l being the mean lever of friction, is

$$2\pi n f N l. \quad\quad\quad\quad (690)$$

(5) *Friction of a Cord on a Cylinder.*—Let R be the radius of the cylinder, s the length of the cord in contact, and T_1 and T_2 the tensions at first and last points of contact, T_2 being greater than T_1. If there were no friction T_1 would be equal to T_2; hence the excess $T_2 - T_1$, when the cord is in a state bordering on motion, is due to friction. Let T be the tensions at the extremities of the elementary arc ds, and let θ be the angle included between their action-lines. Their resultant N is then

$$N = \sqrt{T^2 + 2TT\cos\theta + T^2} = T\sqrt{2(1 + \cos\theta)} = 2T\cos\tfrac{1}{2}\theta. \quad (691)$$

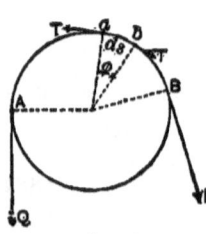

Fig. 69.

If ϕ, Fig. 69, be the angle at the centre of the cylinder subtended by ds, we have

$$\cos \tfrac{1}{2}\theta = \sin \tfrac{1}{2}\phi.$$

Whence, since ϕ is very small,

$$N = 2T \sin \tfrac{1}{2}\phi = T\phi = T\dfrac{ds}{R}. \quad (692)$$

The friction due to N is then

$$fN = fT\dfrac{ds}{R}, \quad\quad\quad\quad (693)$$

and this being the increment of the tension, we have

$$dT = fT\dfrac{ds}{R}. \quad\quad\quad\quad (694)$$

Whence

$$\frac{dT}{T} = f\frac{ds}{R}. \quad \ldots \ldots \ldots (695)$$

Integrating, we have

$$\log T = \frac{fs}{R} + \log C, \quad \ldots \ldots (696)$$

or

$$T = Ce^{\frac{fs}{R}}. \quad \ldots \ldots \ldots (697)$$

When $s = 0$ we have $T = T_1$, and when $s = s$ we have $T = T_2$; therefore

$$T_2 = T_1 e^{\frac{fs}{R}} = T_1 e^{f2\pi n}, \quad \ldots \ldots (698)$$

n being the number of times the cord is wrapped around the cylinder. This relation may be written

$$\frac{T_2}{T_1} = e^{f2\pi n}. \quad \ldots \ldots \ldots (699)$$

From which it is seen that as the number of turns increases in arithmetical progression T_2 increases in geometrical progression. We see, also, how it is possible for a man exerting a tension T_1 on the free end of a rope wound several times around a pile to hold in equilibrium the very much greater tension T_2 of the other end caused by the stoppage of a boat at a wharf.

Stiffness of Cordage.

197. In theoretical mechanics a *cord* is defined to be a collection of molecules so united as to form a perfectly flexible material line. Considering it also to be without weight and to be inextensible, the effect of a force is supposed to be transmitted along its length without loss.

The *tension* of a cord is the intensity of the force which tends to separate any two of its adjacent sections.

Cordage is a term applied to all varieties of lines, cord, and rope, formed by twisting together the textile fibres of hemp, flax, cotton, etc. Since these fibres are neither perfectly flexible nor inextensible themselves, cordage must be much less so, and hence will offer a resistance to being bent from the direction which it naturally assumes. By *stiffness of cordage* is meant the resistance which it offers when it is forced to take a curved form in adapting itself to the surfaces of wheels and pulleys.

The law of this resistance has been deduced by Coulomb from numerous experiments made on different kinds of cordage. He found that the stiffness of cordage is composed of two parts, viz., one, a constant depending on the natural torsion of its fibres; the other, a variable depending on the intensity of the stretching force applied to the cord. He also found that for the same cord it varies inversely with the diameter of the wheel around which the cord is bent. If S be the stiffness, K the constant part due to the natural torsion of the fibres, I that due to a unit of tension, W the total tension, and D the diameter of the wheel, Coulomb's experimental law for a particular cord is expressed by the formula

$$S = \frac{K + IW}{D}. \qquad \ldots \ldots \quad (700)$$

The quantities K and I, according to Morin, ought properly to be expressed in terms of the number n of yarns of which the rope is composed. Making use of Coulomb's results, Morin found that if I be assumed to vary with n, and if K be taken to consist of two terms, one proportional to n and the other to n^2, the values of S derived from

$$S = \frac{n}{D}(0.002148 + 0.001772n + 0.001191\,W) \quad . \quad (701)$$

would conform to all of Coulomb's results for ordinary new *white rope*, and

$$S = \frac{n}{D}(0.01054 + 0.0025n + 0.001372\,W) \quad . \quad . \quad . \quad (702)$$

to those for *tarred rope*. These formulas of Morin are identically those of Coulomb, when for white rope we place

$$K = n(0.002148 + 0.001772n) \quad \text{and} \quad I = n(0.001191),$$

and for tarred rope

$$K = n(0.01054 + 0.0025n) \quad \text{and} \quad I = n(0.00137?).$$

The values of S in pounds, for both kinds of rope, bent over a wheel or axle one foot in diameter, under a tension of one pound, are given in Table VI. For other axles and tensions these values substituted in the above formulas will give the desired results.

The stiffness of partly worn or oily rope is less than that of new rope; it may be only one half as great. The "natural stiffness" K of wet rope is twice that of dry, while the value of I is the same for both.

The stiffness of cordage consumes work when the cord is being wound on the wheel so as to adapt itself to the circumference. This work is equal to the product of the intensity of the resistance by the path described by its point of application, estimated in the direction of the resistance. The path is evidently equal to the length of that portion of the cord wound on the wheel, which is the actual distance passed over by the resistance. Then if N be the number of turns of the wheel, the cord wound is in length equal to $n2\pi R$, and the quantity of work consumed by the stiffness of the cordage will be, for new white rope,

$$Q = \frac{K + IW}{2R} N2\pi R$$
$$= (K + IW)N\pi. \quad . \quad . \quad . \quad . \quad (703)$$

From this we see that the work consumed in N revolutions is independent of the radius of the wheel; as it should be, since the increased stiffness for a wheel of smaller radius is compensated by the less path over which the resistance works in making one complete turn.

Machines.

198. A machine is any instrument or device designed to receive energy from some source, and to overcome certain resistances in transferring this energy to other bodies. Every machine consists of three essential parts, viz., the *driving* point, the *working* point, and the *train*. The first is the point at which the energy is received, the second that at which the transmitted energy is applied, and the third is the series of parts connecting the first and second.

The operating forces in a machine are classified as Powers and Resistances. A *power* is a force which increases or tends to increase the momentum of the parts of a machine; a *resistance* is a force which diminishes or tends to diminish their momentum. Those resistances which the machine is primarily designed to overcome are called *useful*, and the energy expended in overcoming them is called *useful work;* all other resistances are called *prejudicial* or *wasteful*, and the energy expended in overcoming them is called *wasteful or lost work*. From these definitions we see that the action-lines of the powers must either coincide, or make acute angles, with their corresponding virtual velocities, and consequently (Art. 68) their elementary quantities of work will be positive; and that the action-lines of the resistances must either be opposite to, or make obtuse angles with, their respective virtual velocities, and hence their elementary quantities of work will be negative. Recalling the fundamental principle that energy can neither be created nor destroyed, we see that in the discussion of a machine it is necessary to ascertain what amount of energy it has received from the source, and how much of this has been exchanged for useful and lost work, and how much still continues in the machine as potential or kinetic energy. The energy received by the machine is generally designated as the *work of the powers;* that which has been distributed by the machine to masses forming no part of the machine is the *work of the resistances*.

199. *Theory of Machines.*—Resuming the Equation of Energy,

$$\Sigma I dp = \Sigma m \frac{d^2 s}{dt^2} ds,$$

we may apply it to any machine by letting P and dp represent the type-symbols of the intensities and projected virtual velocities of the powers, Q and dq those of the resistances, and m the mass of any particle of the machine; then we have

$$\Sigma P dp - \Sigma Q dq = \Sigma m \frac{d^2 s}{dt^2} ds. \quad \ldots \quad (704)$$

Integrating, we have

$$\Sigma \int P dp - \Sigma \int Q dq = \tfrac{1}{2} \Sigma m v^2 + C \quad \ldots \quad (705)$$

for the general equation of energy applied to machines. If v_s be the type-symbol of the velocities of the masses m at the instant the machine begins to receive energy from the source, we have, since the work of the powers and resistances is then zero,

$$C = -\tfrac{1}{2}\Sigma mv_0^2. \qquad \ldots \qquad (706)$$

Substituting this value of C in the general equation, we have

$$\Sigma \int P dp - \Sigma \int Q dq = \tfrac{1}{2}\Sigma mv^2 - \tfrac{1}{2}\Sigma mv_0^2. \qquad (707)$$

If the machine start from rest, then $v_0 = 0$, and Eq. (707) becomes

$$\Sigma \int P dp - \Sigma \int Q dq = \tfrac{1}{2}\Sigma mv^2. \qquad \ldots \qquad (708)$$

If the integration be taken between any limits corresponding to the states 1 and 2, Eq. (705) will become

$$\Sigma \int_1^2 P dp - \Sigma \int_1^2 Q dq = \tfrac{1}{2}\Sigma mv_2^2 - \tfrac{1}{2}\Sigma mv_1^2. \qquad (709)$$

The theory of machines is embodied in the general equation (707), and may be derived from either of its special forms, Eq. (708) or (709). Taking (709), which applies to any machine in operation, we see that if its kinetic energy increase between any two successive states, the increment is exactly equal to the excess of the work of the powers over that of the resistances in the intervening interval of time. If its kinetic energy diminish, then the loss is equal to the excess of the work of the resistances over that of the powers, and should this condition continue, the machine will come to rest when the total kinetic energy is wholly absorbed in making good the deficiency. If there be no change in the kinetic energy the total work of the powers is exactly equal to that of the resistances during the interval in which the kinetic energy is invariable.

Hence, it appears that when the work of the powers, in any interval whatever, exceeds that of the resistances, the excess is stored up as kinetic energy in the masses which constitute the machine; and when more work is required by the resistances than is supplied by the powers in a given interval of time, the deficiency is made good by the withdrawal of kinetic energy from the parts of the machine. The total quantity of work done by the powers from the instant the machine starts from rest until it comes to rest again, or from any particular state of motion to the same state again, is precisely equal to the work of the resistances in this interval. Therefore, whatever energy is received by the machine is employed in making resistances perform work, and the object of any machine is to get as much useful work done by the expenditure of a definite quantity of energy as is possible. Whatever kinetic energy remains in the machine at any instant is simply the work of the powers which has not heretofore been used in overcoming recurring resistances, and hence is continually accumulating, to be afterwards utilized as necessity requires.

200. *Use of Fly-wheel.*—Due to the construction and application of many machines, it often happens that the energy received from the source, and that consumed by the resistances, vary with the time; and, in addition, these variations of supply and demand may neither be equal nor simultaneous. In such cases, if the machine be of relatively small mass the acceleration of velocity of its parts will be correspondingly great, and the whole machine will be subject to rapid changes of motion, which are often detrimental. To obviate such a defect, the mass of the machine may be increased by the addition of a *fly-wheel*. This consists of a mass of matter distributed in the form of a ring and suitably connected with the rotating shaft on which it is mounted. We have seen that the kinetic energy of rotation is measured by $\frac{1}{2}\omega^2 \Sigma mr^2$, in which ω is the angular velocity and Σmr^2 the moment of inertia of the rotating mass with respect to the axis of rotation. Hence the changes in ω, due to any change in kinetic energy, may be made as small as we please by suitably increasing Σmr^2; this may be done by increasing either the mass or the radius of gyration with respect to the axis. By the introduction of a suitable fly-wheel the changes of velocity may thus be diminished to any desired degree. The greater the moment of inertia of the fly-wheel, the greater will be the quantity of work which it will store up for a given increase in its angular velocity, and, similarly, the more it will yield for a given decrease. As the change of level in a reservoir, due to the addition or discharge of a given quantity of water, will be less noticeable as the surface area is greater, so likewise will be the changes of velocity in the moving parts of a machine, due to a difference between the work of the powers and that of the resistances, according as the moment of inertia of its fly-wheel is greater. From this analogy, the fly-wheel may be regarded as a reservoir of work in a machine.

201. *Efficiency*.—If W be the whole amount of energy supplied to a machine in a given time, during which its kinetic energy remains constant, and W_u and W_l be that employed in overcoming the useful and wasteful resistances respectively, in the same time, then

$$W = W_u + W_l \quad \ldots \ldots \ldots \quad (710)$$

The ratio of the total work to that of the useful resistances, or $\dfrac{W_u}{W}$, is called the *modulus* or *efficiency* of the machine. This ratio is evidently always less than unity, which is its maximum limit, and which it can never reach, since W_l can never in actual machines become zero.

W_l can be diminished in value:

(1) By avoiding all *unnecessary* friction.

(2) By diminishing the intensity of the *necessary* friction; this may be accomplished by selecting material for the contact surfaces whose friction-coefficients are small, or reducing these coefficients by the application of lubricants.

(3) By decreasing the *moments of friction* in rotating parts,

either by decreasing the coefficients as above, or by shortening the lever-arm of friction, or both.

(4) By such an assemblage of parts, arrangement of supports and solidity of foundation, as to avoid sudden and unnecessary vibrations and shocks. These consume work which is dissipated in the form of heat-energy.

The principal sources of energy, whether from fuel, air in motion, animal or water power, etc., have received their supply either directly or indirectly from the sun, which is constantly parting with a portion of its energy in the form of heat. In estimating energy the foot-pound has been assumed as the unit. This unit does not take into consideration the *time* in which the quantity of work is expended, and since the element of time is important in the use of machines, a different unit from the foot-pound is required in measuring their efficiency. Such a unit is the *horse-power*, which corresponds to the expenditure of 550 foot-pounds of work in a second of time. An engine of ten horse-power is one which is capable of doing 5500 foot-pounds of work in a second.

Simple Machines.

202. The *Simple Machines*, or, as they are sometimes called, the *Mechanical Powers*, are the *Cord*, *Lever*, *Pulley*, *Wheel and Axle*, *Inclined Plane*, *Wedge*, and *Screw*. All other machines are formed of combinations of these, and when the relations existing between the powers and resistances are known in the simple machines, the corresponding relations in compound machines may be derived. There is said to be a *gain of power* in a machine when the intensity of the power is less than that of the corresponding useful resistance with which it is compared; and a *loss of power* when the intensity of the power is greater than that of the resistance. These terms are technical, and are used merely to compare the intensities of the powers and resistances, and not to compare the work done by these forces. If we take the quantity of work done by the power and by the useful resistance to be equal, as in the limiting case, it is evident that when there is a gain of power the path described by its point of application, estimated along its action-line, must be greater than that of the resistance; that is, its velocity will be greater than that of the point of application of the resistance, since these unequal distances are described in the same time; this is technically called a *loss of speed*. When there is a loss of power there will be a corresponding *gain of speed*.

The object of discussing Simple Machines is to find the relation existing between the *intensity of the power* and *that of the useful resistance*. To do this, we first place the work of the powers equal to the sum of the works done by all the resistances; the equation so formed we know to be true, whilst the parts of the machine have uniform motion, for during this time the energy received by the machine is wholly absorbed by the work of the resistances.

This condition will always be presupposed, and Eq. (709) will then take the form

$$\Sigma \int_1^2 P dp = \Sigma \int_1^2 Q dq. \quad \ldots \quad (711)$$

We are then often able to eliminate the path factor, and get an equation from which the desired relation between the *intensities* may be obtained. Passing then to the theoretically perfect machine by supposing all the wasteful resistances to be neglected, we find the theoretical ratio of the intensities of the power and useful resistance.

In the following discussion dp and dq are taken to be the projected elementary paths of the points of application of P and Q on the action-lines of these forces respectively, in the time dt, during which the forces are supposed to remain constant in intensity; and ds is the path described by a point at a unit's distance from the axis of rotation in the same time.

The Lever.

203. The *lever* is any solid bar, straight or curved, capable of rotating about a fixed point or line under the action of a power. The point or axis of rotation is called the *fulcrum*. Let AB, Fig. 70, be the axis of a lever, O the axis of the trunnions supporting it, P and Q the power and resistance, p and q their lever-arms with respect to O, and r the radius of the trunnion. The resistances, omitting that of the air, are the useful resistance Q, and the wasteful resistance friction on trunnions, whose intensity will be designated by F. To find F, let θ be the angle included between the action-lines of P and Q; then if N be the intensity of the resultant pressure on the trunnions, we have

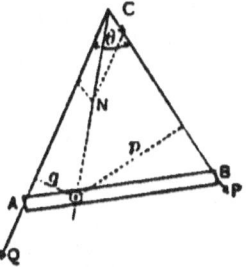

Fig. 70.

$$N = \sqrt{P^2 + Q^2 + 2PQ\cos\theta} \quad \ldots \ldots (712)$$

and

$$F = N\frac{f}{\sqrt{1+f^2}} \quad \ldots \ldots (713)$$

The action-line of N passes through C, the intersection of the action-lines of P and Q, and through the point of contact of the trunnion with the fixed support. Hence CN is the action-line of the resultant pressure. The lever-arm of friction on trunnions is r; therefore the elementary work consumed by friction is

$$Frds; \quad \ldots \ldots \ldots (714)$$

that absorbed by Q is

$$Q d\vec{q} = Qqds; \quad \ldots \ldots \ldots (715)$$

and that done by the power P is

$$Pdp = Ppds. \quad \ldots \ldots \quad (716)$$

Placing the elementary work done by the power equal to the sum of the works consumed by the resistances, under the supposition that the motion of the lever is uniform, we have, after omitting the common factor ds,

$$Pp = Qq + Fr. \quad \ldots \ldots \quad (717)$$

Whence

$$\frac{P}{Q} = \frac{q}{p} + \frac{F}{Q} \cdot \frac{r}{p}. \quad \ldots \ldots \quad (718)$$

Therefore the ratio $\frac{P}{Q}$ of the lever can be found when the quantities P, q, p, r, f and θ are known.

In practice both factors of the last term of Eq. (718), $\frac{F}{Q}$ and $\frac{r}{p}$, are much less than unity, and in ordinary cases their product is negligible. The limit of the ratio of the power to the resistance is the reciprocal of the ratio of their lever-arms; or

$$\frac{P}{Q} = \frac{q}{p}. \quad \ldots \ldots \quad (719)$$

Either P or Q may be taken as the power or resistance, but, to agree with the convention established heretofore, the power is taken to be that force whose virtual moment is positive. When $p > q$ there is a gain of power, and when $p < q$ there is a loss of power.

If the power or resistance, or both, be applied in a plane oblique to the axis of the trunnion, the forces must be resolved into components parallel and perpendicular to the axis. The perpendicular components will replace P and Q in the above discussion, and those which are parallel to the axis will either cause motion in the direction of the axis or produce pressure on the side supports, giving rise to sliding friction, which can readily be computed. The work consumed by this friction will appear among those of the resistances in the equation of equilibrium.

Levers are commonly divided into three classes or orders. In those of the first class the fulcrum is between the power and resistance; in the second the resistance acts between the fulcrum and the power; and in the third the power is applied between the fulcrum and the resistance. As there is no difference in principle in these orders this classification is unimportant.

204. The principles of the lever are involved in the construction of the common balance. To find the conditions of equilibrium, let O, Fig. 71, be the point of suspension, G the centre of gravity of the balance unloaded, $AC = CB = a$, $OC = c$, $OG = h$. Let the balance be loaded with unequal weights $Q > P$, and suppose that it has taken its position of equilibrium as in the figure. Then the moments of the forces about O must be equal; whence we have

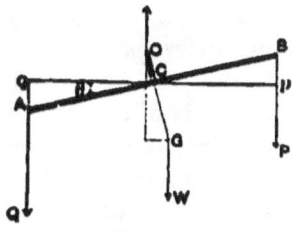

Fig. 71.

$$Q(a \cos \theta - c \sin \theta) = P(a \cos \theta + c \sin \theta) + wh \sin \theta,$$

and therefore

$$\tan \theta = \frac{(Q - P)a}{c(Q + P) + wh}. \quad \ldots \quad (720)$$

The conditions required in a balance are (1) *horizontality* of the beam when the arms are equal in length and the weights In the scale-pans are equal; (2) *sensibility*, which is estimated by the value of θ for a given difference in the weights; (3) *stability*, or the tendency to return to horizontality when the weights are equal. The balance is so constructed that the first condition is satisfied when $Q = P$. The second depends on the greater or less value of θ for a small value of $Q - P$. Assuming a constant difference $Q - P$, we see that tan θ will increase, and hence θ also,

(1) when a is large; that is, when the arms are long;

(2) when c is small, that is, by moving the point of suspension nearer the beam;

(3) when $P + Q$ is small, or the sum of the weights small;

(4) when w, the weight of the balance, is small;

(5) when h is small; that is, when G is not far below the beam.

The sensibility of the balance may be very great and the balance have no stability; that is, no tendency to return to its primitive position after the removal of the weights, and the former will have to be modified to satisfy the latter condition, which is of course essential.

The stability increases with OG, and the sensibility decreases as OG increases. In any particular case the conditions of stability and sensibility must be determined by the uses for which the balance is designed. Thus for rapid weighing of large masses, where great accuracy is not important, the stability must be great; while for the weighing of the precious metals, drugs in small quantities, etc., great sensibility is of primary importance. There is generally some device attached to the balance to check oscillations when the stability is slight.

The Wheel and Axle.

205. This machine consists of a wheel W, Fig. 72, firmly attached to a cylinder C, whose trunnion-ends t and t' rest on trunnion-beds. The power P may be applied through the intervention of a rope passing over a groove in the wheel, or by a crank, or by capstan-bars; its action-line is generally tangent to the circumference of the wheel. The useful resistance Q is applied tangentially to the cylinder C by means of a rope wound upon the cylinder. The principal resistances are Q, and the two wasteful resist-

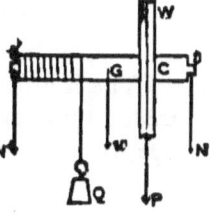

F G 72.

ances, the *stiffness of cordage* caused by Q, and the *friction on trunnions*. Let R, r and ρ be the radii of the wheel, the cylinder and the trunnions, respectively; then the elementary work of the power is

$$Pdp = PRds; \qquad \qquad (721)$$

that of Q is

$$Qdq = Qrds, \qquad \qquad (722)$$

that of stiffness of cordage is

$$\frac{K + IQ}{2r} rds = \frac{1}{2}(K + IQ)ds, \qquad (723)$$

the proper values of K and I being assumed for the kind of rope used; that of friction on trunnions is

$$f'(N + N')\rho ds, \qquad \qquad (724)$$

in which N and N' are the pressures at t and t' due to w, the weight of the machine, and to the forces P and Q; and f' is a symbol for $\dfrac{f}{\sqrt{1+f^2}}$. Placing the work of the power equal to the sum of the works of the resistances, and omitting the common factor ds, we have

$$PR = Qr + \tfrac{1}{2}(K+IQ) + f'(N+N')\rho. \quad . \quad . \quad (725)$$

The limiting ratio of the power to the useful resistance, obtained by neglecting friction on trunnions and stiffness of cordage, is

$$\frac{P}{Q} = \frac{r}{R}, \quad . \quad . \quad . \quad . \quad . \quad . \quad . \quad (726)$$

the same as in the lever; as it should be, since the principle of the two machines is essentially the same.

To find the resultant pressures N and N' at t and t', let l be the length of the axis between the middle points of the trunnions, at which N and N' may be supposed applied; let a, b and c be the distances, estimated parallel to the axis, from the point of application of N to the action-lines of Q, w and P, respectively; the corresponding distances of N' will be $l-a$, $l-b$ and $l-c$. Let the action-lines of w and Q be vertical, and let that of P make an angle ϕ with the vertical. The components of P will then be $P \cos \phi$, vertical, and $P \sin \phi$, horizontal. By the principles of parallel forces, the components of the vertical forces Q, w and $P \cos \phi$, at t, will be

$$Q\frac{l-a}{l}, \quad w\frac{l-b}{l} \quad \text{and} \quad P\frac{l-c}{l} \cos \phi,$$

and the component of $P \sin \phi$ at t will be

$$P\frac{l-c}{l} \sin \phi;$$

and hence the resultant N will be

$$N = \frac{1}{l}\sqrt{[Q(l-a)+w(l-b)+P(l-c)\cos\phi]^2 + P^2(l-c)^2\sin^2\phi}. \quad (727)$$

Similarly the vertical components of N' at l' will be

$$Q\frac{a}{l}, \quad w\frac{b}{l} \quad \text{and} \quad P\frac{c}{l}\cos\phi;$$

and the horizontal component will be

$$P\frac{c}{l}\sin\phi.$$

Whence

$$N' = \frac{1}{l}\sqrt{(Qa + wb + Pc\cos\phi)^2 + P^2c^2\sin^2\phi}. \quad (728)$$

If P be vertical, $\phi = 0$, and we have

$$\left.\begin{aligned} N &= \frac{1}{l}[Q(l-a) + w(l-b) + P(l-c)]; \\ N' &= \frac{1}{l}(Qa + wb + Pc); \end{aligned}\right\} \quad (729)$$

whence

$$N + N' = w + Q + P. \quad \ldots \quad \ldots \quad (730)$$

By substituting the values of N and N' in Eq. (725) the value of the ratio $\frac{P}{Q}$ may be found.

206. *The Differential Wheel and Axle.*—Eq. (726) shows that the gain of power increases as r diminishes; but since, owing to the stress of Q, r cannot be made very small without the liability of bending or breaking the axle, there is a practical limit to the gain of power in the ordinary wheel and axle. In the *differential wheel*, Fig. 73, the axle consists of two cylinders of different radii. The resistance is attached to a movable pulley, and a continuous cord is wound oppositely on the two cylinders of the axle, after partially enveloping the pulley. Supposing the power to be applied tangentially to a wheel, whose radius is R, mounted on the axle, we will have by the equality of moments, neglecting the wasteful resistances, and calling r and r' the radii of the cylinders,

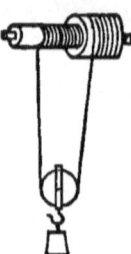

FIG. 73.

$$PR = \tfrac{1}{2}Q(r - r'), \quad \ldots \ldots \quad (731)$$

or

$$\frac{P}{Q} = \frac{r - r'}{2R}. \quad \ldots \ldots \ldots \quad (732)$$

Hence if r' be made great enough to withstand any possible stress of Q, the difference $r - r'$ may be made as small as we please, and thus give any desired gain of power. A combination similar in principle is used to lift very heavy projectiles to the muzzle of the gun.

The Pulleys.

207. (1) *The Fixed Pulley.*—This consists essentially of a grooved wheel supported on trunnions about which it can turn freely, friction being disregarded. The power and useful resistance are applied at the ends of the same cord, which partly envelops the wheel, and is prevented from slipping by the friction between the cord and wheel. If in the wheel and axle the radius of the cylinder be taken the same as that of the wheel, it becomes essentially a fixed pulley, and therefore we may make the equation deduced for the former applicable to the latter by making the proper changes in the quantities which enter it. Hence, for the fixed pulley, Fig. 74, since

$$R = r, \text{ and } a = b = c = \tfrac{1}{2}l,$$

Eq. (727) becomes, taking Q vertical,

$$N = N' = \tfrac{1}{2}\sqrt{(w + Q + P\cos\phi)^2 + P^2\sin^2\phi}, \quad (733)$$

Fig. 74.

From Eq. (725),

$$PR = QR + \tfrac{1}{2}(K + IQ) + f'{}_2N\rho; \quad \dots \quad (734)$$

whence

$$P = Q + \frac{K + IQ}{2R} + f'\frac{\rho}{R}2N. \quad \dots \quad (735)$$

Neglecting stiffness of cordage and friction on trunnions, the limiting ratio of the power to the useful resistance becomes, Eq. (735),

$$\frac{P}{Q} = 1. \quad \dots \dots \dots \quad (736)$$

Theoretically, then, the least value of the power is equal to the useful resistance, and the fixed pulley is used simply to change the direction of the action-line of the power.

The factor $f'\dfrac{\rho}{R}$ of the last term, Eq. (735), is generally very small, and a sufficiently approximate value for this term may be obtained by putting Q for P in the value of N, and omitting w, the weight of the pulley. This being done, we have, Eq. (733),

$$N = \tfrac{1}{2}Q\sqrt{2(1 + \cos\phi)} = Q\cos\tfrac{1}{2}\phi. \quad . \quad . \quad (737)$$

Let θ be the arc of the wheel enveloped by the cord; then

$$\cos\tfrac{1}{2}\phi = \sin\tfrac{1}{2}\theta,$$

and

$$N = Q\sin\tfrac{1}{2}\theta. \quad . \quad . \quad . \quad . \quad . \quad . \quad (738)$$

Substituting this in Eq. (735), we have

$$P = Q\left(1 + 2f'\frac{\rho}{R}\sin\tfrac{1}{2}\theta\right) + \frac{K + IQ}{2R} \quad . \quad . \quad . \quad (739)$$

for the nearly exact relation between the power and useful resistance in the fixed pulley.

If P and Q be parallel, $\theta = 180°$, and Eq. (739) becomes

$$P = Q\left(1 + 2f'\frac{\rho}{R}\right) + \frac{K + IQ}{2R}$$

$$= Q\left(\frac{R + 2f'\rho + \tfrac{1}{2}I}{R}\right) + \frac{K}{2R}; \quad . \quad . \quad (740)$$

which, since the coefficient of Q and the second term are constant for the same rope and pulley, may be written

$$P = \alpha + \beta Q. \quad . \quad \quad . \quad . \quad . \quad (741)$$

208. (2) *The Movable Pulley* (Fig. 75).— When one end of the cord is attached to a fixed point, and the useful resistance W is connected directly with the pulley, the latter is called a *movable pulley*. The resistance in this case is equal to $2N$, and if the weight w of the pulley be neglected, we have, Eq. (738),

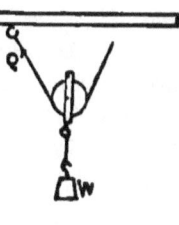

Fig. 75.

$$N = \tfrac{1}{2} W = Q \sin \tfrac{1}{2}\theta;$$

whence

$$Q = \frac{W}{2 \sin \tfrac{1}{2}\theta}$$

Substituting this value for Q in Eq. (739), we have

$$P = W\left(\frac{1}{2 \sin \tfrac{1}{2}\theta} + f\frac{\rho}{R}\right) + \frac{K + l\dfrac{W}{2 \sin \tfrac{1}{2}\theta}}{2R}, \quad \ldots \quad (742)$$

or, since θ is usually 180°,

$$P = W\left(\tfrac{1}{2} + f\frac{\rho}{R}\right) + \frac{K + \tfrac{1}{2} l W}{2R}, \quad \ldots \ldots \quad (743)$$

for the relation between the power P and the useful resistance W, in the movable pulley.

If stiffness of cordage and friction be neglected, the limiting ratio, obtained from Eq. (742), is

$$W = P_2 \sin \tfrac{1}{2}\theta = P\frac{C}{R}; \quad \ldots \quad (744)$$

in which C is the chord of the arc enveloped by the rope. Therefore, neglecting wasteful resistances, in the movable pulley *the power is to the useful resistance as the radius of the pulley is to the chord of the enveloped arc.* When the arc is between 0° and 60°, and between 300° and 360°, there is a loss of power; and when between 60° and 300°, there is a gain of power; the greatest gain of power is at 180° and the greatest loss at 0° or 360°.

209. (3) *The Block and Fall.*—A set of two blocks of pulleys, connected by a continuous cord, arranged to pass alternately from a pulley of one block to one of the other as in Fig. 76, is called a *block and fall*, or a *block and tackle*. One of the blocks is attached to a fixed point, and the other to the useful resistance W. To find the relation between P and W, let $t_1, t_2, t_3, \ldots t_n$, and P, be the successive tensions on the straight portions of the cord, t_1 being that on the first portion of the cord which is attached to one of the blocks; let R be the radius of each pulley, and ρ that of each pulley trunnion, and take $\tfrac{1}{2}\theta$ to be 90°. Then the relations between the successive tensions will be given by Eq. (741), in which Q will be in succession t_1, t_2, etc., and P will be successively t_2, t_3, t_4, etc., and α and β are easily determined constants depending on the rope and pulleys. Then we have

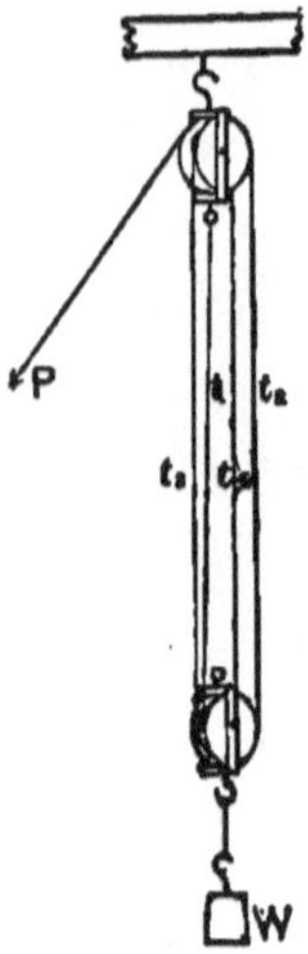

FIG. 76.

$$\left.\begin{aligned}
t_2 &= \alpha + \beta t_1 &&= \alpha \frac{1-\beta}{1-\beta} + \beta t_1; \\
t_3 &= \alpha + \beta t_2 &&= \alpha \frac{1-\beta^2}{1-\beta} + \beta^2 t_1; \\
&\quad\cdot \qquad\qquad\cdot \\
t_n &= \alpha + \beta t_{n-1} &&= \alpha \frac{1-\beta^{n-1}}{1-\beta} + \beta^{n-1} t_1; \\
P &= \alpha + \beta t_n &&= \alpha \frac{1-\beta^n}{1-\beta} + \beta^n t_1;
\end{aligned}\right\} \quad (745)$$

and also

THE PULLEYS.

$$W = t_1 + t_2 + t_3 + \ldots t_n$$
$$= \alpha \frac{(n-1-\beta-\beta^2\ldots-\beta^{n-1})}{1-\beta} + (1+\beta+\beta^2+\ldots\beta^{n-1})t_1$$
$$= \alpha\left(\frac{n}{1-\beta} - \frac{1-\beta^n}{(1-\beta)^2}\right) + \frac{1-\beta^n}{1-\beta}t_1. \quad \ldots \quad (746)$$

Whence we have

$$t_1 = W\frac{1-\beta}{1-\beta^n} - \alpha\left(\frac{n}{1-\beta^n} - \frac{1}{1-\beta}\right). \quad \ldots \quad (747)$$

Substituting this value of t_1 in the expression for t_n, Eqs. (745), we have

$$t_n = W\frac{(\beta-1)\beta^{n-1}}{\beta^n-1} + \alpha\left(\frac{n\beta^{n-1}}{\beta^n-1} - \frac{1}{\beta-1}\right); \quad \ldots \quad (748)$$

which substituted in the last of Eqs. (745) gives

$$P = t_{n+1} = W\frac{(\beta-1)\beta^n}{\beta^n-1} + \alpha\left(\frac{n\beta^n}{\beta^n-1} - \frac{1}{\beta-1}\right). \quad (749)$$

In the case illustrated in the figure $n = 4$, and hence

$$P = W\frac{(\beta-1)\beta^4}{\beta^4-1} + \alpha\left(\frac{4\beta^4}{\beta^4-1} - \frac{1}{\beta-1}\right). \quad \ldots \quad (750)$$

If stiffness of cordage and friction be neglected, then $\alpha = 0$ and $\beta = 1$, and we have

$$\frac{P}{W} = \frac{1}{n}; \quad \ldots \quad \ldots \quad (751)$$

that is, the limiting ratio of the power to the useful resistance is equal to the reciprocal of the number of the parallel portions of the cord which support the resistance.

210. (4) *Other Combinations of Fixed and Movable Pulleys.*—The value of the limiting ratio of the power to the useful resistance depends only on the *number of movable pulleys*, the *arrangement of the cord*, and the *method of attaching it to the resistance;* an inspection of the combination is generally sufficient to establish the required relation. Thus if n, Fig. 77, be the number of pulleys in the first combination, this ratio is readily seen to be

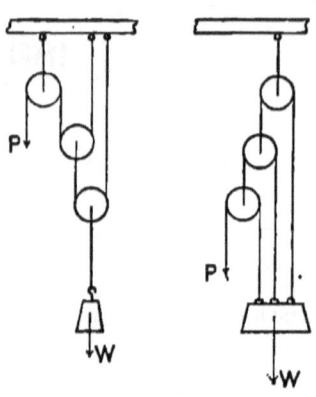

Fig. 77.

$$\frac{P}{W} = \frac{1}{2^n - 1}; \quad \cdots \quad (752)$$

and in the second combination,

$$\frac{P}{W} = \frac{1}{2^n - 1}. \quad \cdots \quad (753)$$

The Inclined Plane.

211. Replacing I in Eqs. (673) and (672) by P' and P'', we have

$$P' = W \frac{\sin i + f \cos i}{\cos \theta + f \sin \theta}; \quad \cdots \quad (754)$$

$$P'' = W \frac{\sin i - f \cos i}{\cos \theta - f \sin \theta}. \quad \cdots \quad (755)$$

Considering the inclined plane as a machine, the first equation expresses the relation of the power P to the resistance of a body's weight W, when the body is either in uniform motion up the plane or in a state bordering on such motion; and the second equation gives the relation, when the body is in uniform motion down the plane or in a state bordering on such motion. The difference of these intensities, for the same weight and angles, is

$$P' - P'' = W \frac{2f \cos (i + \theta)}{\cos^2 \theta - f^2 \sin^2 \theta}; \quad \ldots \quad (756)$$

hence, if the body be in a state bordering on motion up the plane, P' may be diminished in intensity by this value before the body reaches the state bordering on motion downward.

Considering W, Fig. 78, as the only resistance, the limiting value of the ratio $\frac{P}{W}$, obtained by making $f = 0$ in either equation, is

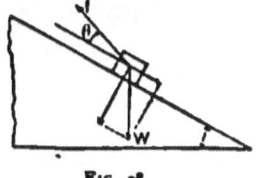

Fig. 78.

$$\frac{P}{W} = \frac{\sin i}{\cos \theta} \quad \ldots \quad (757)$$

When θ is zero the power acts parallel to the plane and upward, and we have

$$\frac{P}{W} = \sin i = \frac{h}{l}; \quad \ldots \quad (758)$$

that is, the power is to the resistance as the height of the plane is to its length, and there is always a gain of power.

When $\theta = 360° - i$ the power acts horizontally, pressing the body against the plane, and we have

$$\frac{P}{W} = \tan i = \frac{h}{b}; \quad \ldots \quad (759)$$

that is, the power is to the resistance as the height of the plane is to its base; and there is a gain of power when the plane has a less inclination than 45°, and a loss of power for greater inclinations.

212. The elementary quantity of work expended by the power P in moving the body uniformly up the plane is, when the action-line of P is parallel to the plane, Eq. (711),

$$Pds = Wds \sin i + fWds \cos i$$
$$= Wdh + fWdl; \quad \ldots \ldots \quad (760)$$

in which dl and dh are the horizontal and vertical projections of ds. Integrating between any limits, we have

$$\int Pds = W(h - h') + fW(l - l'); \quad \ldots \quad (761)$$

or, the total quantity of work is equal to the work stored as potential energy of the weight, plus the work consumed by friction due to the weight over a path equal to the horizontal projection of the actual path of the body.

The Wedge.

213. The *wedge* usually consists of a solid triangular prism, as ABC (Fig. 79), which is inserted into an opening between two bodies or parts of the same body, to split or separate them. The surface AB, to which the pressure or blow is given, is called the *back;* the surfaces AC and BC, the *faces;* and their line of intersection C, the *edge* of the wedge.

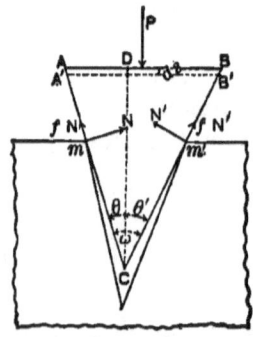

Fig. 79.

Let the wedge ABC be inserted within the jaws of the opening, and be in contact with them along lines projected in m and m'. Suppose the normal pressures N and N' and the force P to lie in the same plane, and the latter to be normal to the back of the wedge. If the wedge move forward, or be in a state bordering on motion forward, the friction between the jaws of the opening and the surface of the wedge, due to the normal pressures N and N', will act from m and m' towards A and B respectively. If the wedge fly back, or be in a state bordering on motion outward, the friction will oppose this motion or tendency, and act along the faces toward C. Considering the first case, and supposing that the wedge is in equilibrium due to the forces acting, we have, for the components parallel to AB,

$$Nf \sin \theta - N \cos \theta + N' \cos \theta' - N'f \sin \theta' = 0; \quad (762)$$

and for those parallel to DC, perpendicular to AB,

$$P - N \sin \theta - Nf \cos \theta - N' \sin \theta' - N''f \cos \theta' = 0. (763)$$

Eliminating N' from these equations, and representing the angle of the wedge by $\omega = \theta + \theta'$, we have

$$N = \frac{P(\cos \theta' - f \sin \theta')}{(1 - f^2) \sin \omega + 2f \cos \omega}; \quad \ldots \quad (764)$$

and similarly eliminating N, we have

$$N' = \frac{P(\cos \theta - f \sin \theta)}{(1 - f^2) \sin \omega + 2f \cos \omega}. \quad \ldots \quad (765)$$

From these values we have

$$P = \frac{N[(1 - f^2) \sin \omega + 2f \cos \omega]}{\cos \theta' - f \sin \theta'}$$
$$= \frac{N'[(1 - f^2) \sin \omega + 2f \cos \omega]}{\cos \theta - f \sin \theta}. \quad \ldots \quad (766)$$

In order that P may have a possible value for a state bordering on motion forward we must have

$$\cos \theta' > f \sin \theta' \quad \text{and} \quad \cos \theta > f \sin \theta,$$

or
$$\cot \theta' > f \quad \text{and} \quad \cot \theta > f. \quad \ldots \quad (767)$$

Substituting for f its value $\tan \alpha$, we have

$$\cot \theta' > \tan \alpha \quad \text{and} \quad \cot \theta > \tan \alpha,$$

or

$$\theta' < 90° - \alpha \quad \text{and} \quad \theta < 90° - \alpha. \quad \ldots \quad (768)$$

Hence

$$\omega < 180° - 2\alpha. \quad \ldots \ldots \quad (769)$$

That is, *in order that the wedge may be driven in, the angle of the wedge must be less than* 180° *diminished by twice the angle of friction.*

If the wedge be in a state bordering on motion outward, the friction terms in Eq. (766) will change their signs, and we have

$$P = \frac{N[(1-f^2)\sin\omega - 2f\cos\omega]}{\cos\theta' + f\sin\theta'}$$
$$= \frac{N'[(1-f^2)\sin\omega - 2f\cos\omega]}{\cos\theta + f\sin\theta}. \quad \ldots \quad (770)$$

If some pressure be required to prevent the wedge from flying out, we must have

$$(1-f^2)\sin\omega > 2f\cos\omega,$$

or

$$\tan\omega > 2\frac{\tan\alpha}{1-\tan^2\alpha}; \quad \ldots \ldots \quad (771)$$

which reduces to

$$\tan\omega > \tan 2\alpha,$$

or

$$\omega > 2\alpha. \quad \ldots \ldots \ldots \quad (772)$$

Hence we see that *in order that the wedge may be held in its place when the external pressure is removed, the angle of the wedge must be less than twice the angle of friction.*

When the wedge is used as a power N and N' are generally nearly equal, and either may be considered as the resistance to be overcome. The particular form of the wedge in any case depends on the special use for which it is intended. Thus in splitting wood it is usually made isosceles, and the power P is applied to the back of the wedge as an impulsion. The axe, chisel, engraver, knife, tool of a plane, and the raised projections of a file, are examples of wedges, whose forms are modified in accord with the above principles, for the particular purposes for which they are designed. Taking the wedge to be isosceles and $N = N'$, we have for the ratio of P to N

$$\frac{P}{N} = \frac{(1 - f^2) \sin \omega + 2f \cos \omega}{\cos \tfrac{1}{2}\omega - f \sin \tfrac{1}{2}\omega}; \quad \ldots \quad (773)$$

and omitting friction,

$$\frac{P}{N} = \frac{\sin \omega}{\cos \tfrac{1}{2}\omega} = 2 \sin \tfrac{1}{2}\omega. \quad \ldots \quad (774)$$

Hence the gain of power increases very rapidly as the angle of the wedge diminishes.

The Screw.

214. The screw combines the principles of the lever and inclined plane. It consists usually of a solid circular cylinder, called the *newel*, on the surface of which is a *thread* or *fillet*, whose section by a plane through the axis of the cylinder is usually either a rectangle or a triangle. The thread of the screw is a volume which may be generated by a rectangle or triangle having its base on the cylindrical surface and always parallel to the axis of the newel, moving uniformly around and along the axis. Every point of the generating area will therefore describe a helix, and the upper and under sides will describe helicoidal surfaces. The distance between the successive positions of the same point of the generating area, measured in the direction of the axis of the newel, after one complete revolution, is called the *pitch* of the screw, or the *helical interval*. In screws with rectangular threads the pitch must be greater than the base of the generating rectangle; in triangular threads it is usually equal to the base of the generating triangle.

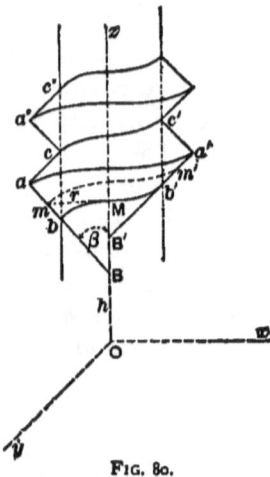

FIG. 80.

The screw is engaged in a nut whose interior cylindrical surface is screw-cut in such a manner as to fit the fillet accurately. The useful resistance to be overcome, if the nut be fixed in position, is applied to the foot of the screw so that its action-line may be in the direction of the axis of the newel; if the nut have freedom of motion and the screw is fixed, the useful resistance is applied to the nut.

Take the axis of the newel as the axis of z, and let abc, Fig. 80, be the generating area. Let

β, be the constant angle made by ab in all of its positions with z.

r, the distance of any helix from the axis, constant for the same helix, but variable for different helices.

γ, the constant angle made by any assumed helix with the horizontal plane.

ϕ, the angle through which the screw or nut is rotated.

l, the lever arm of the power.

For the elementary work of the power we have

$$Pdp = Pld\phi, \qquad \qquad (775)$$

and for the work of the useful resistance, Q,

$$Qds = Qrd\phi \tan \gamma = Qr'd\phi \tan \gamma', \quad \ldots \quad (776)$$

in which r' is the radius of a mean helix, and γ' the angle which this helix makes with the horizontal.

Let f be the coefficient of friction, and N the normal pressure, and suppose the friction concentrated on the mean helix whose radius is r'. Then we have for the elementary work of friction

$$\frac{fNr'd\phi}{\cos \gamma'}, \quad \ldots \quad \ldots \quad (777)$$

Hence

$$Pld\phi = Qr'd\phi \tan \gamma' + \frac{fNr'd\phi}{\cos \gamma'}, \quad \ldots \quad (778)$$

or

$$P = \frac{r' \tan \gamma'}{l}\left(Q + \frac{fN}{\sin \gamma'}\right). \quad \ldots \quad (779)$$

215. To find the relation between P and Q it is necessary to find N in terms of Q. From the equilibrium of the forces we know that the algebraic sum of their intensities in any direction is equal to zero. Hence, resolving the forces P, Q, N and fN, we have, for the sum of their components in the direction of the axis z,

$$Q + fN \sin \gamma' - N \cos \theta_z = 0, \quad \ldots \quad (780)$$

in which θ_z is the angle made by the normal to the helicoidal surface, at the assumed point of equilibrium on the mean helix, with the axis of z. The cosine of this angle is

$$\cos \theta_z = \frac{1}{\sqrt{1 + \tan^2 \gamma' + \cot^2 \beta}}; \quad \ldots \quad (781)$$

and Eq. (780) becomeses

$$Q = N(\cos\theta_z - f\sin\gamma'), \quad \ldots \quad (782)$$

or

$$N = \frac{Q}{\cos\theta_z - f\sin\gamma'}. \quad \ldots \ldots \quad (783)$$

Substituting this value of N in Eq. (779), we have

$$P = \frac{Qr'\tan\gamma'}{l}\left(1 + \frac{f}{\sin\gamma'(\cos\theta_z - f\sin\gamma')}\right). \quad \ldots \quad (784)$$

Replacing $\cos\theta_z$ by its value, and reducing, we have

$$P = \frac{Qr'\tan\gamma'}{l}\left(1 + \frac{f\sqrt{1 + \tan^2\gamma' + \cot^2\beta}}{\sin\gamma' - f\sin^2\gamma'\sqrt{1 + \tan^2\gamma' + \cot^2\beta}}\right)(785)$$

If the thread be rectangular, $\beta = 90°$, and we have

$$P = \frac{Qr'\tan\gamma'}{l}\left(1 + \frac{f\sqrt{1 + \tan^2\gamma'}}{\sin\gamma' - f\sin^2\gamma'\sqrt{1 + \tan^2\gamma'}}\right); \quad \ldots \quad (786)$$

hence, all other things being equal, the screw with a rectangular thread is more advantageous than one with a triangular thread.

If we suppose the friction to be neglected, then $f = 0$, and the limiting ratio of the power to the resistance is

$$\frac{P}{Q} = \frac{r\tan\gamma}{l}. \quad \ldots \ldots \ldots \quad (787)$$

Multiplying and dividing the second member by 2π, we have

$$\frac{P}{Q} = \frac{2\pi r\tan\gamma}{2\pi l}; \quad \ldots \ldots \quad (788)$$

or *when friction is neglected the power is to the useful resistance as the helical interval is to the circumference described by the extremity of the lever arm of the power;* hence there is usually a great gain of power in the application of this machine.

The Cord.

216. Let a perfectly flexible and inextensible cord assume a position of equilibrium under the action of any forces whatever. The resultant of the forces acting at either extremity must be in the direction of the cord at that extremity; for, if it have a component perpendicular to the cord, the latter, being perfectly flexible, must move in the direction indicated by the perpendic-

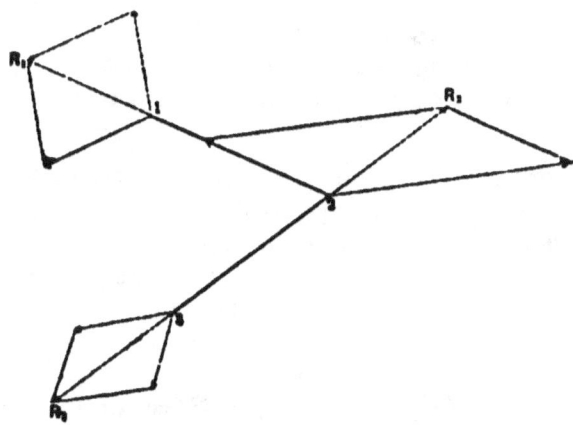

Fig. 81.

ular component. Let 1 2 3, Fig. 81, be the cord, the resultant R_1 being in the direction 2 1. Since the point of application of a force may be taken to be any point of its action-line within the limits of the body on which it acts, the resultant R_1 may be considered as applied at the point 2. Then the resultant of all the forces acting at 2, including R_1, must be in the direction 3 2; and this resultant, R_2, may be considered as applied at the point 3.

Thus in any case each successive resultant may have its point of application transferred until all the action-lines have a common point, and the conditions of equilibrium will be the same as before. Therefore *the conditions of equilibrium for a perfectly flexible and inextensible cord, under the action of any forces whatever, are the same as if all the forces were applied at a single point, their intensities and directions remaining unchanged.*

217. Let T_0 be the tension of the cord at the origin, assumed at any point; I, the type-symbol of the intensities of the extraneous forces; T, the tension at any point; and let θ_x, θ_y, θ_z; α, β, γ; ϕ_x, ϕ_y, ϕ_z, be the angles which T_0, I and T make with the co-ordinate axes, respectively. Then from the equilibrium of the system we have

$$\left. \begin{array}{l} T\cos\phi_x = T_0\cos\theta_x + \Sigma I\cos\alpha; \\ T\cos\phi_y = T_0\cos\theta_y + \Sigma I\cos\beta; \\ T\cos\phi_z = T_0\cos\theta_z + \Sigma I\cos\gamma; \end{array} \right\} \quad \ldots \quad (789)$$

the last terms comprising all the extraneous forces between the origin and the point where the tension is T.

If forces act at all points of the cord in such a manner as to make the tension vary by continuity, then the cord will assume the form of a curve, and Eqs. (789) become

$$\left. \begin{array}{l} T\dfrac{dx}{ds} = T_0\cos\theta_x + \Sigma I\cos\alpha; \\ T\dfrac{dy}{ds} = T_0\cos\theta_y + \Sigma I\cos\beta; \\ T\dfrac{dz}{ds} = T_0\cos\theta_z + \Sigma I\cos\gamma. \end{array} \right\} \quad \ldots \quad (790)$$

Such a curve is called a *funicular curve*, and Eqs. (790) are its differential equations.

If the extraneous forces be parallel and coplanar, we may

assume the curve in the plane xs, with the forces parallel to s, and we have

$$\cos \alpha = \cos \alpha'' = \text{etc.} = 0;$$

hence

$$T\frac{dx}{ds} = T_{\bullet} \cos \theta_x = \text{a constant;} \quad \ldots \quad (791)$$

that is, *the component of the tension perpendicular to the direction of the forces and in their plane is constant.*

218. Let the cord $a\ b\ c\ d$, Fig. 82, be in equilibrium, *the length of each branch representing its own tension, and the tensions being assumed constant throughout.* Let R_1 and R_2 be equal and opposite

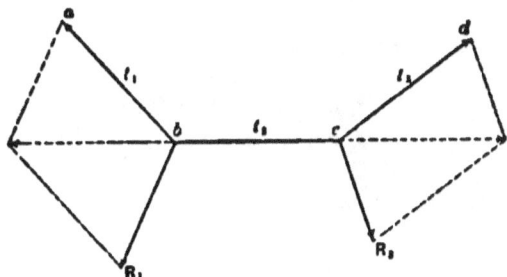

Fig. 82.

to the resultants of the tensions acting at the points b and c, and let the symbols $(t_1\ t_2)$, $(t_1\ R_1)$, etc., represent the angles made by the corresponding lines in the figure. When three forces acting at a single point are in equilibrio, their intensities are directly as the sines of the opposite angles, and we have

$$\frac{t_1}{\sin (t_2 R_1)} = \frac{t_2}{\sin (t_1 R_1)} = \frac{R_1}{\sin (t_1 t_2)}; \quad \ldots \quad (792)$$

$$\frac{t_2}{\sin (t_3 R_2)} = \frac{t_3}{\sin (t_2 R_2)} = \frac{R_2}{\sin (t_2 t_3)}; \quad \ldots \quad (793)$$

and since the tensions are equal,

$$\left. \begin{array}{l} \sin(t_2 R_1) = \sin(t_1 R_1); \\ \sin(t_2 R_2) = \sin(t_2 R_2). \end{array} \right\} \quad \cdots \quad (794)$$

That is, *when the tensions are equal throughout, the resultant of the forces at any point bisects the angle made by the adjacent branches of the cord.*

219. Let circumferences be passed through each vertex and the two adjacent ones, and denote their radii by r_1 and r_2, and let s be the length of one branch of the cord. Then

$$\left. \begin{array}{l} r_1 \cos \tfrac{1}{2}(t_1 t_2) = \tfrac{1}{2}s; \\ r_2 \cos \tfrac{1}{2}(t_2 t_3) = \tfrac{1}{2}s. \end{array} \right\} \quad \cdots \quad (795)$$

We have also, from the figure,

$$\left. \begin{array}{l} t_2 \cos \tfrac{1}{2}(t_1 t_2) = \tfrac{1}{2}R_1; \\ t_2 \cos \tfrac{1}{2}(t_2 t_3) = \tfrac{1}{2}R_2. \end{array} \right\} \quad \cdots \quad (796)$$

From the latter equations we have

$$\frac{R_1}{\cos \tfrac{1}{2}(t_1 t_2)} = \frac{R_2}{\cos \tfrac{1}{2}(t_2 t_3)}, \quad \cdots \quad (797)$$

which by Eqs. (795) reduce to

$$r_1 R_1 = r_2 R_2. \quad \cdots \quad (798)$$

That is, *the intensities of the resultants are inversely as the radii of the circumferences passing through their points of application and the two adjacent vertices.*

220. From Arts. 218 and 219 we conclude that when the *funicular curve* has a constant tension throughout, *the resultants of the forces acting at the different points are normal to the curve, and their intensities vary inversely as the radii of curvature at their points of application.*

The Catenary Curve.

221. The curve assumed by a heavy flexible and inextensible cord under the action of its own weight is called a *catenary curve*. Assume the curve in the plane xz, and we have for its differential equations

$$T\frac{dx}{ds} = T_{\circ}\cos\theta_x = c; \quad \ldots \quad \ldots \quad (799)$$

$$T\frac{dz}{ds} = \int g\delta\omega ds + T_{\circ}\cos\theta_z; \quad \ldots \quad (800)$$

in which δ, ω and ds are the density, cross-section and length of an elementary portion of the curve.

From Eq. (799) we see that *the horizontal component of the tension is constant;* and since at the lowest point $\frac{dx}{ds} = 1$, *the horizontal component of the tension at any point is equal to the tension at the lowest point.*

Taking the origin at the lowest point, we have $T_{\circ}\cos\theta_z = 0$, and from Eq. (800) we have

$$T\frac{dz}{ds} = g\int_0^s \delta\omega ds. \quad \ldots \quad \ldots \quad (801)$$

That is, *the vertical component of the tension at any point is equal to the weight of that portion of the cord between this point and the lowest point.*

Having the vertical and horizontal components, the tension is readily constructed.

222. *The Common Catenary* (Fig. 83).—When δ and ω are constant the curve is called the *common catenary*, and we have

$$T\frac{dz}{ds} = g\delta\omega s; \quad \ldots \quad \ldots \quad (802)$$

or, making $g\delta\omega = 1$, which introduces the condition that the unit of length of the cord gives a unit of weight, we have

$$T \cdot \frac{dz}{ds} = s. \quad . \quad . \quad (803)$$

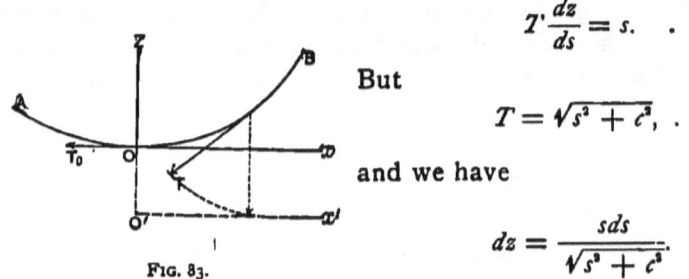

FIG. 83.

But
$$T = \sqrt{s^2 + c^2}, \quad . \quad . \quad (804)$$

and we have
$$dz = \frac{s\,ds}{\sqrt{s^2 + c^2}}. \quad . \quad (805)$$

Integrating, making $s = 0$ when $z = 0$, we have

$$z + c = \sqrt{s^2 + c^2}, \quad . \quad . \quad . \quad . \quad . \quad (806)$$

from which we get

$$s^2 = z^2 + 2cz. \quad . \quad . \quad . \quad . \quad . \quad . \quad (807)$$

From Eq. (799) we have

$$dx = \frac{c\,ds}{T} = \frac{c\,ds}{\sqrt{s^2 + c^2}}, \quad . \quad . \quad . \quad . \quad (808)$$

and integrating, as before,

$$x = c \log \left[\frac{s}{c} + \sqrt{1 + \frac{s^2}{c^2}} \right]$$

$$= c \log \left[\frac{\sqrt{z^2 + 2cz}}{c} + \sqrt{1 + \frac{z^2 + 2cz}{c^2}} \right]$$

$$= c \log \frac{\sqrt{z^2 + 2cz} + z + c}{c}, \quad . \quad . \quad . \quad . \quad . \quad (809)$$

which is the equation of the common catenary.

223. Substituting the value of s^2 from Eq. (807) in Eq. (804), we have

$$T = \sqrt{z^2 + 2cz + c^2} = z + c. \quad . \quad . \quad . \quad (810)$$

The tension at any point is therefore given by the ordinate of the curve estimated from a right line parallel to the axis of x and at a distance below the origin equal to c. This line is called the *directrix* of the curve; it is readily constructed either from the tension at any point or from the constant horizontal component of the tension.

When the cord is vertical the directrix passes through the lowest point and is perpendicular to the cord, and when the cord is horizontal the directrix is at an infinite distance below the cord and parallel to it.

MECHANICS OF FLUIDS.

Introductory.

224. *Fluids* are either *compressible* or *incompressible*. Gases are called compressible fluids because they readily change their volume with a change of extraneous pressure. Liquids are said to be incompressible, because their change of volume for even very great changes of pressure is practically inappreciable.

225. In the following discussions the fluids considered are supposed to be *perfect ;* that is, to possess *perfect elasticity of volume* and *perfect mobility*. The latter condition requires the fluid to be without *viscosity*, or each molecule to be a *free rigid solid ;* but this freedom of the molecule of course refers only to its motion *in the fluid*, as the containing vessel necessarily exerts a constraining effect on the molecules in contact with it.

226. *The stress at any point of a fluid is measured by the pressure per unit area at that point.*

The *compression* is measured by the ratio of the change in volume to the original volume, or $\frac{dv}{v}$.

The *compressibility* is measured by the ratio of the compression to the stress which produces it, or $\frac{dv}{v} \div dp$.

The *elasticity of volume* (Art. 22) is measured by the ratio of the stress to the compression which it produces, or $dp \div \frac{dv}{v}$; the compressibility and elasticity of volume are therefore reciprocals of each other.

227. Equal Transmission of Fluid Pressure; Pascal's Principle. —When a fluid is in equilibrium, the resultant pressure on the surface of the enclosing vessel at any point is normal to the surface, and the pressure on each molecule of the fluid, neglecting its weight, is the same in all directions. Therefore *any extraneous pressure applied to a fluid in equilibrium is transmitted equally in all directions.*

Laws of Perfect Gases.

228. Action of Heat on Gases.—*Heat* is a form of energy, and it can be measured either in *units of heat* or *units of work*. A *unit of heat* is the *quantity of heat* required to raise the temperature of a *unit weight* of water from the standard temperature to one degree above that temperature. When the pound is the unit weight the equivalent quantity of work is about 778.8 foot-pounds* when the thermal unit is taken for the F. scale at 60°, and about 1402 foot-pounds for the C. scale at the equal temperature 15°.5. This is called the Mechanical Equivalent of Heat, or Joule's Equivalent, and is represented by the symbol J.

The heat of a substance is due to its molecular kinetic energy, and a given body at a given temperature contains a definite, measurable quantity of heat.

When a gas is contained in a vessel the pressure on the sides of the vessel is supposed to be due to the heat in the gas; that is, the molecules have a certain average velocity, and the continuous succession of an exceedingly great number of impacts of the molecule against the sides of the vessel gives rise to a sustained pressure.

The *specific heat* of a substance is the number of units of heat required to raise the temperature of a *unit weight* of the substance from the standard temperature to one degree above that temperature. This quantity of heat will evidently vary with the conditions under which it is added to the substance. It is called *specific heat at constant volume* when the substance does not

---

* Rowland's Mechanical Equivalent of Heat, 1880.

change its volume during the application of heat, and *specific heat at constant pressure* when the pressure remains the same.

The *specific volume* of a substance is the volume of a unit weight of the substance.

The *specific gravity* is the weight of a unit volume of the substance; the specific volume and specific gravity are therefore reciprocals of each other.

229. When a gas expands it does work in overcoming the resistance due to the pressure of the envelope. Therefore if a gas be allowed to expand while heat is added to it, the pressure remaining constant, part of the heat will be used in doing the work of expansion. At the same time the temperature of the gas will rise; for the number of impacts per unit of surface will *decrease*, thus requiring the intensity of each impact to *increase* in order to keep the pressure constant. It is found by experiment that the quantity of heat added to the gas is greater* than the sum of the quantities required to do the extraneous work and raise the temperature. The remaining energy is supposed to be taken up in overcoming attractions between the molecules, that is, in doing *internal work*. This quantity of work is small and not well determined, and it decreases as the gas approaches the condition of a perfect gas; it is not, therefore, directly discussed. But the sum of this quantity, and the quantity of energy in the gas as heat, is assumed to be constant for a given condition of the gas. Call this sum *internal energy*, and represent it by U. Let W represent the *external work*, and suppose a quantity of heat represented by dQ to be added to the gas. Then we shall have

$$dQ = dU + dW. \quad \ldots \ldots \quad (811)$$

If we suppose heat to be added to or withdrawn from the body, so as to cause it to pass through a series of changes, and

---

* In the case of hydrogen the quantity is less. This would indicate a molecular repulsion in hydrogen, while in other gases the molecular forces are attractions. (See Atom; Encyclopedia Britannica, vol. iii., 9th edition.)

then finally be brought back to its original condition, we will have

$$\int_2^1 dQ = U_1 - U_2 + \int_2^1 dW = \int_2^1 dW, \quad . \quad . \quad (812)$$

since in this case $U_1 - U_2 = 0$. Such a series of changes is called a *cycle*. Therefore we conclude that, when a body has undergone a cycle of changes due to the application of heat, the total quantity of heat transferred has been wholly converted into external work.

Let a cylinder contain a gas whose pressure per unit of area is p and volume v, and let it be closed by a piston of area a, capable of working without friction. Then, if h be the altitude of the volume, the elementary quantity of work done in any change of volume is

$$padh = pdv = dW,$$

and Eq. (811) becomes

$$dQ = dU + pdv. \quad . \quad . \quad . \quad . \quad . \quad (813)$$

The simplest illustration of this general principle is found in the case of gases, where the external work of expansion can be determined from the change in volume alone; p being taken as constant throughout the change.

230. *Laws of the Gaseous State.*—The experimental laws which have been found applicable to a gas are:

(1) *Boyle's or Mariotte's Law.*—This law asserts that if a given mass of gas be kept at a *constant temperature*, the product of its volume and pressure per unit area is always constant. If, therefore, v_0 and v be the specific volumes corresponding to any particular temperature, and p_0 and p be the corresponding pressures, Boyle's law is expressed by the equation

$$pv = p_0 v_0 = \text{a constant}. \quad . \quad . \quad . \quad (814)$$

Since $v = \dfrac{1}{\delta g}$, this law may also be expressed, in terms of pressure and density, by

$$p\delta_o = p_o\delta. \qquad \qquad (815)$$

Differentiating Eq. (814), we have

$$pdv + vdp = 0; \qquad \qquad (816)$$

whence

$$-dp\frac{v}{dv} = p; \qquad \qquad (817)$$

or, *the elasticity of a gas obeying Boyle's law is measured by the pressure per unit of area.*

(2) *Law of Charles or Gay Lussac.*—If a given mass of gas be subjected to a constant pressure, then its temperature and volume will vary under the law expressed by the equation

$$v = v_o(1 + \alpha t); \qquad \qquad (818)$$

in which v_o and v are the specific volumes of the gas at the temperature $0°$ and $t°$ respectively, and α is the constant coefficient of expansion for the particular gas and the assumed thermometric scale. Hence the law of Charles asserts *that the volume of a gas subjected to a constant pressure is equal to the volume at the standard temperature increased by a constant fractional part of this volume for each increase of $1°$ temperature.* For atmospheric air and the Centigrade scale this constant has been found to be .003665 = $\frac{1}{273}$; and as it differs but little from this value for other permanent gases, α is taken to be the same constant for all permanent gases. If the volume be kept constant, the pressure and temperature will vary according to the law expressed by

$$p = p_o(1 + \alpha t), \qquad \qquad (819)$$

which is another expression of the law of Charles, in which α has the same numerical value as before.

231. Boyle's and Charles's Laws combined.—When the pressure is constant we have from Charles's law

$$v = v_0(1 + \alpha t).$$

Multiplying by the constant pressure p_0, we have

$$p_0 v = p_0 v_0 (1 + \alpha t). \quad \ldots \ldots \quad (820)$$

By Boyle's law for the temperature t we have

$$p_0 v = p_1 v_1, \quad \ldots \ldots \ldots \quad (821)$$

in which p_1 and v_1 are the pressure and corresponding specific volume at the temperature t.

Hence we have, from Eqs. (821) and (820),

$$p_1 v_1 = p_0 v_0 (1 + \alpha t); \quad \ldots \ldots \quad (822)$$

in which p_1, v_1, are the pressure and specific volume of the gas at the temperature t, and p_0, v_0, are the pressure and volume at the temperature 0° C. Hence the general law connecting the pressure, temperature, and specific volume of a permanent gas is, *that the product of the pressure and specific volume at any temperature is equal to the corresponding product at 0° C., increased by $\frac{1}{273}$ of its value for each degree above 0° C., and correspondingly diminished for each degree below 0° C.*

This law of a perfect gas is only approximately true for ordinary gases.

232. Using the Centigrade scale, Eq. (822) may be written

$$pv = p_0 v_0 \frac{273 + t}{273} = \frac{p_0 v_0 \tau}{273} = R\tau, \quad \ldots \quad (823)$$

in which τ is temperature measured on the Centigrade scale, from an origin 273° below 0° C., and R, a constant, is, Eq. (813), the work of expansion at constant pressure corresponding to a change of 1° on that scale. This origin is called the *absolute zero* of temperature, and temperatures measured on such a scale are called *absolute temperatures*. Hence, Eq. (823), pressures under a constant volume, or volumes under a constant pressure, vary directly with the absolute temperature of the gas. Also, since, when τ is zero, the specific volume and pressure are zero, we may infer that, were such a temperature possible, the matter would no longer be in a gaseous state, but behave as a solid.

233. *Differential Equations of the Specific Heats.*—Resuming the fundamental equation (813),

$$dQ = dU + pdv,$$

and combining it with Eq. (823), we can readily derive the differential equation connecting the pressure, specific volume and absolute temperature of a gas with the quantity of work corresponding to the change in its quantity of heat. Since the condition of a perfect gas is completely determined when any two of the three quantities p, v and τ are known, the required equation may be expressed in three forms.

(1) Let τ and v be the independent variables; then we have

$$dU = \frac{dU}{d\tau} d\tau + \frac{dU}{dv} dv. \quad \ldots \ldots \quad (824)$$

Experiments made by Regnault indicated that $\dfrac{dU}{dv}$ is zero, but the more accurate experiments of Joule show that it has an appreciable value for actual gases. Hence we may assume for the perfect gas

$$dU = \frac{dU}{d\tau} d\tau, \quad \ldots \ldots \ldots \quad (825)$$

and from Eq. (813) we have

$$dQ = \frac{dU}{d\tau} d\tau + pdv; \quad \ldots \ldots \quad (826)$$

or, since $\frac{dU}{d\tau}$ is the specific heat at constant volume, we have, denoting this by C_v,

$$dQ = C_v d\tau + pdv = C_v d\tau + R\frac{\tau}{v} dv. \quad \ldots \quad (827)$$

(2) Let τ and p be the independent variables. Then differentiating Eq. (823), we have

$$pdv + vdp = Rd\tau, \quad \ldots \ldots \quad (828)$$

whence

$$pdv = Rd\tau - vdp,$$

$$= Rd\tau - \frac{R\tau}{p} dp. \quad \ldots \ldots \quad (829)$$

Substituting this value for pdv in Eq. (827), we have

$$dQ = (C_v + R) d\tau - \frac{R\tau}{p} dp; \quad \ldots \ldots \quad (830)$$

or, since $(C_v + R)$ is the specific heat at constant pressure, we have, denoting it by C_p,

$$dQ = C_p d\tau - \frac{R\tau}{p} dp. \quad \ldots \ldots \quad (831)$$

(3) Let p and v be the independent variables; then substituting in Eq. (827) the value of $d\tau$ derived from Eq. (828), we have

$$dQ = \frac{C_v}{R} vdp + \frac{C_p}{R} pdv. \quad \ldots \ldots \quad (832)$$

Eqs. (827), (831) and (832) may then be written, in terms of the specific heats,

$$dQ = C_v d\tau + (C_p - C_v)\frac{\tau}{v} dv, \quad \ldots \quad (833)$$

$$dQ = C_p d\tau + (C_v - C_p)\frac{\tau}{p} dp, \quad \ldots \quad (834)$$

$$dQ = \frac{C_v}{C_p - C_v} v dp + \frac{C_p}{C_p - C_v} p dv, \quad \ldots \quad (835)$$

which are different expressions for the same relations between p, v and τ. These, it will be remembered, are expressed in units of work, and their equivalent values in heat-units can be obtained by dividing by the constant J.

234. Poisson's Laws.—After a gas has been subjected to a compression or an expansion so sudden as to prevent any heat from being transferred to or from it through the envelope, then, Q being a constant, dQ will be zero. Then, p and v being the independent variables, Eq. (835) reduces to

$$\frac{C_v}{C_p - C_v} v dp + \frac{C_p}{C_p - C_v} p dv = 0, \quad \ldots \quad (836)$$

whence

$$\frac{C_p}{C_v} + \frac{v dp}{p dv} = 0; \quad \ldots \quad (837)$$

which becomes, when the ratio of the specific heats is represented by k,

$$k \frac{dv}{v} + \frac{dp}{p} = 0. \quad \ldots \quad (838)$$

Integrating between any two limits, p_0, v_0, and p_1, v_1, we have

$$k \log \frac{v_1}{v_0} = \log \frac{p_0}{p_1} = \log \left(\frac{v_1}{v_0}\right)^k; \quad \ldots \quad (839)$$

and passing to the corresponding numbers, we have

$$\frac{p_0}{p_1} = \left(\frac{v_1}{v_0}\right)^k, \quad \ldots \quad (840)$$

or, in terms of densities,

$$\frac{p_1}{p_0} = \left(\frac{\delta_1}{\delta_0}\right)^k. \quad \ldots \ldots \quad (841)$$

Taking τ and v as variables, Eq. (833), under the same supposition, gives

$$C_v d\tau + (C_p - C_v)\frac{\tau}{v} dv = 0. \quad \ldots \ldots \quad (842)$$

Dividing by τC_v, we have

$$\frac{d\tau}{\tau} + (k - 1)\frac{dv}{v} = 0. \quad \ldots \ldots \quad (843)$$

Integrating between limits τ_0, v_0, and τ_1, v_1, we have

$$\log\frac{\tau_1}{\tau_0} = (k - 1)\log\frac{v_0}{v_1} = \log\left(\frac{v_0}{v_1}\right)^{k-1}. \quad \ldots \quad (844)$$

Whence we have

$$\frac{\tau_1}{\tau_0} = \left(\frac{v_0}{v_1}\right)^{k-1}, \quad \ldots \ldots \quad (845)$$

or, in terms of corresponding densities,

$$\frac{\tau_1}{\tau_0} = \left(\frac{\delta_1}{\delta_0}\right)^{k-1}. \quad \ldots \ldots \quad (846)$$

Equations (840) and (845) express the laws of Poisson and give the relations existing between the pressure, specific volume and absolute temperature of a gas subjected to sudden expansion or compression, under the supposition that the gas neither receives nor parts with heat.

235. The law of the elastic force can be found from the above equations by obtaining the value of the elasticity, $-dp\frac{v}{dv}$. Thus, from Eq. (838), we have

$$dp = -k\frac{pdv}{v}, \quad \ldots \ldots \ldots (847)$$

and therefore

$$-dp\frac{v}{dv} = kp; \quad \ldots \ldots \ldots (848)$$

that is, *the elasticity of a gas suddenly compressed or expanded is equal to the elasticity under the law of Boyle multiplied by the ratio of its specific heats.* This ratio for all permanent gases is very nearly constant and equal to 1.41.

236. *Properties of Actual Gases.*—The laws of Boyle, Charles and Poisson are applicable to the theoretically perfect gas and are only approximately true for actual gases. For example, Regnault found by experiment that the compressibility of air constantly increased with the pressure within the limits of his experiments, and is greater than that exacted by Boyle's law; that nitrogen also has a greater compressibility than the theoretical, although less than that of air; that hydrogen conforms very nearly to theory, but its compressibility diminishes with increasing pressure; and that carbon dioxide departs from the law much more than the other named gases, even under very small pressures. More recently Cailletet has shown that at a pressure of 80 atmospheres air has a maximum relative compressibility. Despretz found that carbon dioxide, sulphuretted hydrogen, ammonia and cyanogen are more compressible than air, while hydrogen is the same as air up to 15 atmospheres, beyond which it is less compressible than air. All gases deviate less from the law as their temperatures increase, but when approaching their temperatures of liquefaction, or critical temperatures, their deviations increase rapidly. Boyle's law is taken to be the governing law connecting pressure and density, under ordinary temperatures and pressures, for all gases that are difficult to liquefy, such as oxygen, hydrogen, nitrogen, air, etc., for the error committed is generally negligible.

Hydrostatics.

237. *Hydrostatics* treats of *fluids at rest*, and *Hydrodynamics* of *fluids in motion*.—The pressure on any surface in a fluid at rest is composed of two parts: one due to transmitted extraneous pressure, and the other to the weight of the fluid. If p' be the extraneous pressure per unit area, and s the area of the surface, we shall have, by the principle of the *equal transmission of pressure*, $p's$ for the transmitted pressure on the area s.

238. To find the second it will be necessary to obtain an expression for the pressure due to weight at any point of the surface, and then integrate between the proper limits for the whole surface considered. The fluid will arrange itself in parallel horizontal homogeneous layers; and, whatever be the boundary of the fluid, the pressure due to weight at the highest point will be zero. Take the horizontal plane of reference, Fig. 84, to be that through the highest point of the fluid, and let the axis of z be vertical and positive downward. Then, p representing the pressure due to weight at any point, we shall have

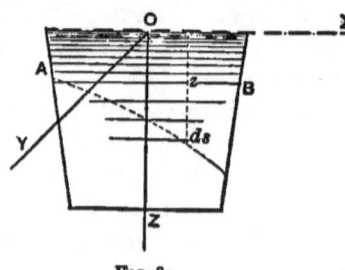

Fig. 84.

$$p = f(z);$$

and if the fluid be homogeneous throughout we shall have, the pressure being directly proportional to the depth,

$$p = \delta g z, \quad \ldots \ldots \ldots \quad (849)$$

in which δ represents the density of the fluid and g the weight of the unit of mass; δg is therefore the specific gravity.

Representing the pressure due to weight on the surface s by P, we have
$$dP = \delta g z ds,$$
whence
$$P = \delta g \int z ds. \quad \ldots \ldots \quad (850)$$

But from the principle of the centre of mass we have
$$\int z ds = s\bar{z};$$

$\bar{z}$ being the vertical co-ordinate of the centre of gravity of the surface s; whence we have
$$P = \delta g s \bar{z}. \quad \ldots \ldots \quad (851)$$

From which we have the following principle: *The pressure on any surface, due to the weight of a homogeneous fluid, is equal to the weight of a prismatic column of the fluid whose base is equal to the area of the surface and whose height is the depth of the centre of gravity of the surface below the upper surface of the fluid.*

239. If α, β, γ, be the angles which the normal to the elementary surface ds makes with rectangular co-ordinate axes x, y, z, the components of dP will be

$$\delta g z \cos\alpha ds, \quad \delta g z \cos\beta ds, \quad \delta g z \cos\gamma ds.$$

But $\cos\alpha ds$, $\cos\beta ds$, $\cos\gamma ds$, are the projections of the elementary area ds on the planes yz, xz and xy, respectively. Hence the component pressure on any elementary area, due to the weight of the fluid, is equal to the weight of a prismatic volume of the fluid whose base is the projection of the area on a plane normal to the direction considered, and whose altitude is the depth of the elementary area below the free surface of the fluid.

Whatever be the area of the surface pressed, it may be supposed divided into elementary areas ds, and the component intensities in any direction on each of these may be found as in the above case. Summing these, we have, for the three components of the whole pressure on the surface,

$\delta g \Sigma z \cos\alpha ds$, $\delta g \Sigma z \cos\beta ds$, $\delta g \Sigma z \cos\gamma ds$,

in the directions of the co-ordinate axes x, y, z, respectively. Since in these expressions z, α, β and γ are in general independent variables, the component pressures cannot always be found.

But if the surface pressed be a plane surface, the cosine factors become constant, and we have

$\delta g \cos\alpha \Sigma z ds$, $\delta g \cos\beta \Sigma z ds$, $\delta g \cos\gamma \Sigma z ds$,

which by the principle of the centre of mass become

$$\delta gs \cos\alpha \bar{z}, \quad \delta gs \cos\beta \bar{z}, \quad \delta gs \cos\gamma \bar{z}. \quad \ldots \quad (852)$$

Therefore *the pressure on any immersed plane surface estimated in any direction is equal to the weight of a prismatic volume of the fluid whose base is equal to the projection of the area on a plane normal to that direction, and whose altitude is the depth of the centre of gravity of the plane area below the free surface.*

240. It is readily seen that the pressure on any immersed surface will remain constant, provided the altitude of the upper surface of the fluid from the centre of gravity of the surface be

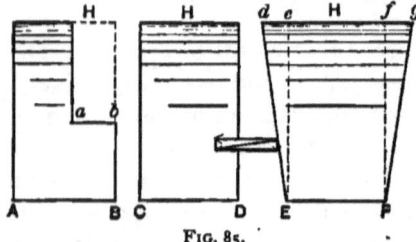
Fig. 85.

unchanged, however much the quantity of the fluid be varied. For example, the pressure on the equal areas AB, CD, EF, Fig. 85, due to the constant weight of the liquid in the several vessels will be the same if the free surface in each be at the same level, although the volumes of the fluid differ considerably. But while the pressures on these equal areas are equal, the resultant vertical pressure is in each case the weight of the fluid contained in the vessel. Thus the vertical pressure upward on ab is equal to the weight of a volume of the fluid equal to abH, and the downward pressure on AB is the weight of the volume

ABH. Hence the resultant pressure is downward and equal to the weight of the liquid actually in the vessel. Similarly in the vessel HEF, the weights of the triangular volumes deE and fgF are added to that of $efEF$ in determining the pressure upon the support which sustains the vessel. This principle is sometimes called the *Hydrostatic Paradox*.

241. *Centre of Pressure.* —It has been shown, Art. 38, that the centre of a system of parallel forces is the point of application of the resultant of the system. The liquid pressures on a *plane surface* form such a system, and their centre is called a *centre of pressure*. To find its position, let M, Fig. 86, be any immersed plane surface, and let AB be its intersection with the free surface of the liquid; let ds represent any elementary area of the plane surface at a, a', etc., s, s', etc., its projection on the free surface of the liquid, and ψ the angle which M makes with the same surface. The pressure on ds is normal to M at a, and is

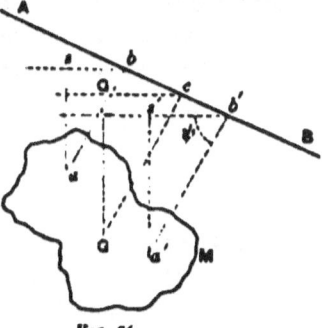

Fig. 86.

$$\delta g s ds = \delta g r \sin\psi\, ds, \quad \ldots \quad \ldots \quad (853)$$

in which r is the distance ab. The moment of this pressure with respect to AB is

$$\delta g r^2 \sin\psi\, ds, \quad . \quad\quad . \quad (854)$$

and the sum of the moments of all the elementary pressures with respect to AB is

$$\delta g \sin\psi \Sigma r^2 ds. \quad \ldots \quad \ldots \quad (855)$$

The total or resultant pressure on M is

$$\delta g \bar{z} A = \delta g h \sin\psi A, \quad \ldots \quad \ldots \quad (856)$$

in which A is the whole area of M, $\bar{z}$ the depth of the centre of gravity of M below the free surface, and h its distance from AB. If l be the distance of the centre of pressure from AB, we shall have, since the moment of the resultant pressure is equal to the sum of the moments of the component pressures with respect to the same line,

$$\delta g h \sin\psi A l = \delta g \sin\psi \Sigma r^2 ds. \quad \ldots \quad (857)$$

Whence

$$l = \frac{\Sigma r^2 ds}{Ah} = \frac{A(k_i^2 + h^2)}{Ah} = \frac{k_i^2 + h^2}{h}. \quad \ldots \quad (858)$$

Therefore the centre of pressure is at the centre of percussion of the surface with respect to the line AB as a spontaneous axis. If the plane surface be rectangular and one of its sides lie in the upper surface of the liquid, then, since $k_i^2 = \frac{h^2}{3}$, we shall have

$$l = \frac{4h}{3}; \quad \ldots \quad (859)$$

that is, the centre of pressure is on the median line two thirds the length of that line from its intersection with the free surface. If the plane surface be horizontal, its centre of pressure coincides with its centre of gravity. In all cases the determination of the centre of pressure depends on finding the value of k_i^2 with respect to a parallel to its intersection with the free surface passing through its centre of mass.

242. In many cases the centre of pressure may be constructed graphically. For example, in the right-angled triangle ABC, Fig. 87, whose base lies in the upper surface of the liquid, the centres of pressure of its elements considered as horizontal right lines will all lie on the right line joining C with the middle point of AB. Similarly, considering its elements to be the right lines parallel to AC, their centres of pressure will lie on the line Bb, drawn from B to a point two thirds the length of AC from A; hence p will be the centre of pressure of the triangle, and it is readily seen to be at a distance from AB equal to one half the depth of the triangle.

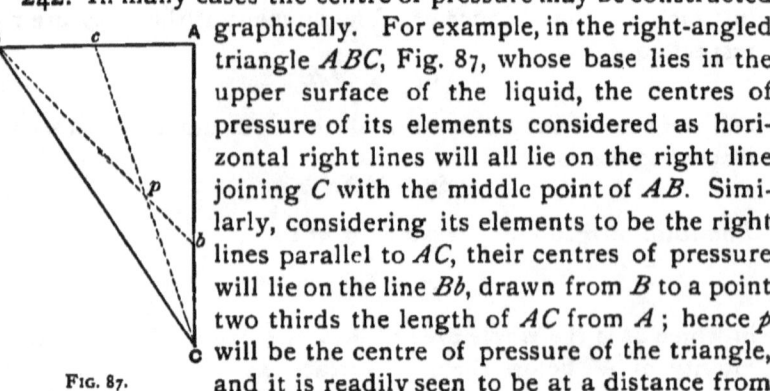

FIG. 87.

243. Buoyant Effort of Fluids.—*When a body is at rest in a fluid, the sum of the component pressures in any horizontal direction is zero.* To show this, let the direction be parallel to the axis of x; then it is to be proved that the expression

$$\delta g \Sigma s \cos a ds \quad \ldots \ldots \quad (860)$$

is the sum of pairs of equal positive and negative products, and consequently becomes zero for any horizontal direction whatever. Conceive the volume of the body M, Fig. 88, to be composed of

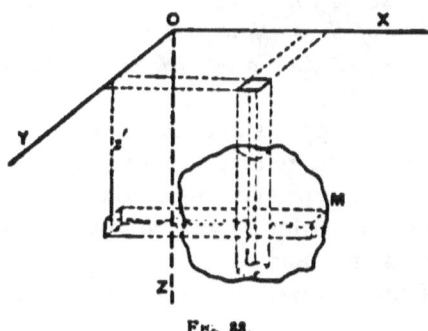

Fig. 88.

elementary horizontal prisms or cylinders parallel to the axis of x. The base, cosads, of any one extended to the plane ys is the projection on ys of opposite surface elements of the body having the same coordinate z. Although the opposite surface elements ds are in general unequal, the angles a are so related that their cosines have contrary signs, and, since the products $ds \cos a$ are numerically equal, we will have

$$\delta g \Sigma s \cos a ds = 0. \quad \ldots \ldots \quad (861)$$

As the axis x may be taken to be any axis in the horizontal plane, we see that the component pressures of the fluid in any horizontal direction counterbalance each other.

Considering now the vertical components of the fluid pressure, represented by

$$\delta g \Sigma z \cos\gamma ds, \quad \ldots \quad \ldots \quad (862)$$

we see that although the factors cos γds are numerically equal and have contrary signs, the coordinate z differs for the upper and lower surface by the height of the filament of the body cut out by the infinitesimal cylinder. Representing these coordinates by z' and z'' respectively, we have

$$\delta g \Sigma z \cos\gamma ds = - \delta g \Sigma (z' - z'') \cos\gamma ds = - \delta g V, \quad (863)$$

in which V represents the volume of the displaced fluid. Therefore, *the resultant of the vertical pressures on the body is equal to the weight of the volume of fluid displaced.* The action-line of this resultant, being parallel to its components, is vertically upward. To find the point of application of this resultant, let $x_{\prime}, y_{\prime}$, be its horizontal coordinates, referred to the centre of gravity of the body, and x', y', the type symbols of the horizontal coordinates of the points of application of the component vertical pressures referred to the same point. Then, by the principle of equality of moments of the resultant and its components, we have

$$\left. \begin{array}{l} \delta g \Sigma (z' - z'') \cos\gamma x' ds = \delta g V x_{\prime}, \\ \delta g \Sigma (z' - z'') \cos\gamma y' ds = \delta g V y_{\prime}; \end{array} \right\} \quad \ldots \quad (864)$$

whence

$$\left. \begin{array}{l} x_{\prime} = \dfrac{\Sigma (z' - z'') \cos\gamma x' ds}{V} = \dfrac{\Sigma v x'}{V}, \\ \\ y_{\prime} = \dfrac{\Sigma (z' - z'') \cos\gamma y' ds}{V} = \dfrac{\Sigma v y'}{V}. \end{array} \right\} \quad \ldots \quad (865)$$

Therefore the point $x_{\prime}, y_{\prime}$ is the centre of gravity of the displaced fluid. This point is called the *centre of buoyancy*, and the resultant vertical pressure is called the *buoyant effort* of the fluid. The principle of the buoyant effort is often called the *principle of Archimedes*.

244. If the density of the body be greater than that of the fluid, it can be at rest only when acted on by another force directed vertically upward and equal in intensity to the difference between the buoyant effort and the weight of the body. If the density of the body be equal to that of the fluid, it will remain at rest at any position where it is wholly submerged. When the density of the body is less than that of the fluid, it will remain at rest when the weight of the fluid displaced is equal to that of the body.

245. The problem of determining the specific gravities of bodies is based on the principle of Archimedes, and the apparatus employed are called Hydrometers, Densimeters, or Areometers. The operation of accurately finding the specific gravity of a body requires great care and attention to many minute details. Practically the body is weighed in air, and its weight in vacuo determined from the known conditions of temperature and barometric pressure. Then its loss of weight in water at $4°$ C. is found, which will be the weight of an equal volume of the standard water. If W be its weight in vacuo at $4°$ C., and w its loss of weight in water at $4°$ C., then we have

$$S = \frac{W}{w}. \qquad \qquad (866)$$

If the temperature of the water be $t°$, and α be the cubic dilatation of the unit volume of water from $4°$ C. to $t°$ C., then the unit volume of water becomes $1 + \alpha$ at $t°$. But the ratio of the densities of water at $4°$ and $t°$ is $1 + \alpha : 1$. Hence if w be the weight of the displaced water at $t°$, its weight at $4°$ C. will be $w(1 + \alpha)$. The specific gravity of the body when the temperature of the water is $t°$ is then

$$S = \frac{W}{w(1 + \alpha)}. \qquad \qquad (867)$$

The values of α for all temperatures from $0°$ C. to $100°$. C. have been carefully determined, and are given in Table VII, Appendix.

246. Stability of Floating Bodies.—If a floating body be slightly displaced from its position of rest, and then return to it after a series of oscillations, under the action of its own weight and the buoyant effort of the liquid, its position of rest is called a position of *stable equilibrium*. To ascertain the conditions of stable equilibrium of a floating body, let, Fig. 89,

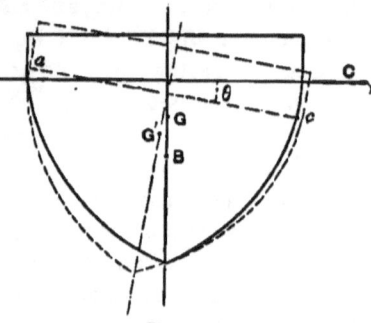

Fig. 89.

AC be the upper surface of the liquid, or plane of flotation.

ac, the original section of the body by the upper surface before displacement.

G, G', the original and any subsequent positions of the centre of gravity of the body.

B, the centre of buoyancy for the position of stable equilibrium.

h, the distance GB; positive when B is below G, and negative when B is above G.

z_0, the distance of G below the plane of flotation.

z, the variable distance of G' from the same plane.

V', V, the volume of the body and the original volume of the displaced liquid, respectively.

δ', δ, the mean density of the body and the density of the liquid, respectively.

θ, the variable angle between the planes AC and ac, whose maximum value is taken to be so small as to permit us to place $\sin \theta = \theta$ and $\cos \theta = 1 - \dfrac{\theta^2}{2}$ without sensible error.

Then will $z_0 \pm h$ be the original distance of B below the plane of flotation, and $z \pm h \cos \theta = z \pm h \left(1 - \dfrac{\theta^2}{2}\right)$ the variable distance of B below the same plane.

If now we suppose the body pushed downward and at the same time slightly careened, the potential energy added to the body in this position may be considered as the work of three forces, viz.: (1) the weight of the body, (2) the buoyant effort of the original displaced liquid, and (3) the buoyant effort of the increment of the displaced liquid.

The potential energy due to the first is

$$\delta' V' g (z - z_0), \quad \ldots \quad \ldots \quad (868)$$

and to the second is

$$\delta V g [z \pm h \cos \theta - (z_0 \pm h)] = \delta V g [(z - z_0) \pm h (1 - \frac{\theta^2}{2} - 1)]$$

$$= \delta V g \left[(z - z_0) \mp h \frac{\theta^2}{2} \right]; \quad . \quad (869)$$

the upper sign corresponding to a positive, and the lower to a negative, value of h. To find the potential energy due to the third force, let s be the area of ac, and ds the elementary area; let z' be the distance of ds from the upper surface of the liquid, and x its distance from the axis of the plane ac passing through the centre of mass of ac; and let I represent the moment of inertia of s with respect to the axis. Then the variable volume of the liquid between the planes AC and ac may be considered to be made up of all the vertical liquid prisms whose upper bases are $ds \cos \theta$, having their lower bases at the distance z' below the free surface AC. The buoyant effort of each of these is represented by the type symbol $z' \delta g ds \cos \theta$, and the corresponding elementary potential energy is

$$z' \delta g ds \cos \theta dz' = \delta g ds z' dz' \text{ nearly}, \quad \ldots \quad (870)$$

The whole potential energy due to the third force is therefore

$$\delta g \int ds \int z' dz' = \delta g \frac{1}{2} \int z'^2 ds. \quad \ldots \quad (871)$$

But we have

$$z' = z_0' \pm x \sin \theta = z_0' \pm x \theta,$$
$$z'^2 = z_0'^2 \pm 2\theta x z_0' + x^2 \theta^2.$$

Hence

$$\delta g \frac{1}{2}\int z''^2 ds = \delta g \frac{1}{2}\left[\int z_0''^2 ds \pm 2\int \theta x z_0' ds + \int x^2 \theta^2 ds\right]$$

$$= \delta g \frac{1}{2}\left(z_0''^2 s + I\theta^2\right); \quad \ldots \ldots \ldots \quad (872)$$

since

$$\theta \int x z_0' ds = \theta z_0' \int x ds = 0,$$

from the principle of the centre of mass.

By the principle of the conservation of energy, Art. 156, the maximum kinetic energy of the body will occur at the instant the body passes the position of stable equilibrium, and be equal to the sum of the potential energies originally transferred to the body by the forces producing displacement. Hence, representing the masses and velocities of the molecules by m and v, we have

$$\frac{1}{2}\Sigma m v^2 = \delta' V' g(z - z_0) - \delta V g\left[(z - z_0) \mp h\frac{\theta^2}{2}\right]$$

$$- \delta g \frac{1}{2}\left(z_0''^2 s + I\theta^2\right), \quad . \quad (873)$$

which, since $\delta' V' g(z - z_0) = \delta V g(z - z_0)$, reduces to

$$\frac{1}{2}\Sigma m v^2 = -\frac{1}{2}\delta g\left[z_0''^2 s + I\theta^2 \mp V h \theta^2\right]$$

$$= -\frac{1}{2}\delta g\left[z_0''^2 s + \theta^2(I \mp V h)\right]. \quad \ldots \quad (874)$$

247. For stable equilibrium the second member must be negative, which requires the quantity within the parenthesis to be positive. The first term, $z_0''^2 s$, is the factor of the potential energy due to the variable submergence of the body, and is always positive whether z_0' be positive or negative. The second term, $\theta^2 (I \mp V h)$, is positive for all *negative values of h;* that is, a floating body is always in a position of stable equilibrium whenever the centre of gravity of the body is below the centre of buoyancy. When the centre of gravity of the body is above the centre of buoyancy, that is, when h is *positive*, stable equilibrium is possible only when the condition

$$I > Vh \quad \text{or} \quad h < \frac{I}{V} \quad \ldots \ldots \quad (875)$$

is fulfilled; that is, for principal axes in the plane of flotation whose moments of inertia are greater than the product of the volume of the displaced fluid by the distance between the centre of gravity of the body and the centre of buoyancy. As the longest principal axis is that corresponding to the least moment of inertia, whenever this condition is fulfilled for this axis it is satisfied for all others. Hence in ships the condition is fixed for the axis parallel to the length of the vessel.

248. The point in which the vertical through the centre of buoyancy intersects the line BG when the body is slightly careened is called the *metacentre*. If the equilibrium be stable, the metacentre must lie above the centre of gravity of the body.

Hydrodynamics.

249. The theoretical discussion of fluid motion is limited to the case of a *continuous fluid* enclosed in boundaries formed by the walls of the containing vessel and the free surface, if there be one. A *stream line* is a continuous line, without abrupt changes of direction, made up of contiguous elements, each element when in motion taking the place of that one which precedes it, without there being at any instant vacant spaces between them; the stream line may also be considered as the path pursued by a fluid element. Under these assumptions, the normal section of a fluid at any instant is the surface containing all the fluid elements which pass at that instant normally through the section at every point.

250. *Euler's Equations of Fluid Motion.*—Since every elementary mass of a fluid has theoretically perfect freedom of motion,

any one may be regarded as a free solid and its motion of translation will be governed by Eqs. (T_m). Hence, m being its mass, and x, y, z, the co-ordinates of its centre of mass, we have

$$\left.\begin{array}{c} \Sigma I \cos \alpha - m \dfrac{d^2 x}{dt^2} = 0, \\[4pt] \Sigma I \cos \beta - m \dfrac{d^2 y}{dt^2} = 0, \\[4pt] \Sigma I \cos \gamma - m \dfrac{d^2 z}{dt^2} = 0. \end{array}\right\} \quad \cdots \quad (876)$$

Let mX, mY, mZ, represent the component intensities of the resultant of the forces applied directly to the molecule. Generally, the only force applied directly to the molecule is its weight. To find the components of the resultant of the transmitted pressures, we have

$$p = f(x, y, z).$$

The changes in the component pressures, from molecule to molecule, are

$$\frac{dp}{dx} dx, \quad \frac{dp}{dy} dy, \quad \frac{dp}{dz} dz,$$

and therefore the components which produce acceleration are

$$-\frac{dp}{dx} dx, \quad -\frac{dp}{dy} dy, \quad -\frac{dp}{dz} dz. \quad \cdots \quad (877)$$

Hence the pressures on the elementary areas perpendicular to the corresponding directions are

$$-\frac{dp}{dx} dx\, dy\, dz, \quad -\frac{dp}{dy} dx\, dy\, dz, \quad -\frac{dp}{dz} dx\, dy\, dz. \quad \cdots \quad (878)$$

Substituting the sums of these components for their respective symbols, Eqs. (876) become

$$\left.\begin{array}{l} mX - \dfrac{dp}{dx}dx\,dy\,dz - m\dfrac{d^2x}{dt^2} = 0, \\[4pt] mY - \dfrac{dp}{dy}dx\,dy\,dz - m\dfrac{d^2y}{dt^2} = 0, \\[4pt] mZ - \dfrac{dp}{dz}dx\,dy\,dz - m\dfrac{d^2z}{dt^2} = 0. \end{array}\right\} \quad \ldots \quad (879)$$

Let the volume of the elementary mass m be represented by $dx\,dy\,dz$, and let δ be the density of the fluid at the point x, y, z; then Eqs. (879) reduce to

$$\left.\begin{array}{l} \dfrac{1}{\delta}\dfrac{dp}{dx} = X - \dfrac{d^2x}{dt^2}, \\[4pt] \dfrac{1}{\delta}\dfrac{dp}{dy} = Y - \dfrac{d^2y}{dt^2}, \\[4pt] \dfrac{1}{\delta}\dfrac{dp}{dz} = Z - \dfrac{d^2z}{dt^2}, \end{array}\right\} \quad \ldots \ldots \quad (880)$$

which are the differential equations of the motion of translation of any fluid element.

The velocities of the fluid element are functions of x, y, z and t. Therefore, if u, v and w be the component velocities in the directions of x, y and z, respectively, we have

$$\left.\begin{array}{l} \dfrac{d^2x}{dt^2} = \dfrac{du}{dt} = \left(\dfrac{du}{dt}\right)\dfrac{dt}{dt} + \dfrac{du}{dx}\dfrac{dx}{dt} + \dfrac{du}{dy}\dfrac{dy}{dt} + \dfrac{du}{dz}\dfrac{dz}{dt}, \\[4pt] \dfrac{d^2y}{dt^2} = \dfrac{dv}{dt} = \left(\dfrac{dv}{dt}\right)\dfrac{dt}{dt} + \dfrac{dv}{dx}\dfrac{dx}{dt} + \dfrac{dv}{dy}\dfrac{dy}{dt} + \dfrac{dv}{dz}\dfrac{dz}{dt}, \\[4pt] \dfrac{d^2z}{dt^2} = \dfrac{dw}{dt} = \left(\dfrac{dw}{dt}\right)\dfrac{dt}{dt} + \dfrac{dw}{dx}\dfrac{dx}{dt} + \dfrac{dw}{dy}\dfrac{dy}{dt} + \dfrac{dw}{dz}\dfrac{dz}{dt}, \end{array}\right\} \quad (881)$$

in which the partial differential coefficients of the component velocities regarded as a function of the time are enclosed in parentheses to distinguish them from the corresponding total differentials. Substituting these values of the accelerations in Eqs. (880), and writing u, v and w for their symbols, we have

$$\left.\begin{aligned}\frac{1}{\delta}\frac{dp}{dx} &= X - \left(\frac{du}{dt}\right) - \frac{du}{dx}u - \frac{du}{dy}v - \frac{du}{dz}w, \\ \frac{1}{\delta}\frac{dp}{dy} &= Y - \left(\frac{dv}{dt}\right) - \frac{dv}{dx}u - \frac{dv}{dy}v - \frac{dv}{dz}w, \\ \frac{1}{\delta}\frac{dp}{dz} &= Z - \left(\frac{dw}{dt}\right) - \frac{dw}{dx}u - \frac{dw}{dy}v - \frac{dw}{dz}w. \end{aligned}\right\} \quad . \quad (882)$$

When the five quantities u, v, w, δ and p can be found in terms of x, y, z and t, the problem is capable of solution. Two other equations connecting these quantities are therefore necessary.

251. *Equation of Continuity.*—The mass m, occupying the volume $dx\,dy\,dz$, being $\delta\,dx\,dy\,dz$ at the time t, will be increased or diminished at the time $t + dt$ by the mass

$$dm = dx\,dy\,dz\frac{d\delta}{dt}dt. \quad \ldots \quad \ldots \quad (883)$$

To find an equivalent expression for this change of mass in the elementary volume, consider the faces of the rectangular parallelopipedon in pairs. Then the mass flowing through the face $dy\,dz$ nearest the origin in the time dt is $\delta\,dy\,dz\,u\,dt$, and the mass flowing out of the opposite face in the same time is

$$dy\,dz\left(\delta u + \frac{d\delta u}{dx}dx\right)dt.$$

The difference of these masses is the increment of mass in the elementary volume due to the flow through these faces, and is

$$-\,dx\,dy\,dz\,dt\frac{d\delta u}{dx}.$$

Similarly for the other two pairs of faces we have

$$- dx\, dy\, dz\, dt \frac{d\delta v}{dy},$$

$$- dx\, dy\, dz\, dt \frac{d\delta w}{dz},$$

and the total increment of mass in the elementary volume in the time dt is

$$- dx\, dy\, dz\, dt\left(\frac{d\delta u}{dx} + \frac{d\delta v}{dy} + \frac{d\delta w}{dz}\right), \quad \ldots \quad (884)$$

which, being placed equal to its equivalent mass (883), gives, after reduction,

$$\frac{d\delta}{dt} + \frac{d\delta u}{dx} + \frac{d\delta v}{dy} + \frac{d\delta w}{dz} = 0; \quad \ldots \quad (885)$$

an equation expressing the *principle of continuity of the fluid:* that is, when a fluid mass filling its boundaries is in motion, *the motion of its elements must be consistent with the law of geometrical continuity.*

Eq. (885) may also be written, after performing the indicated differentiations,

$$\frac{d\delta}{dt} + \frac{d\delta}{dx}u + \frac{d\delta}{dy}v + \frac{d\delta}{dz}w + \delta\left(\frac{du}{dx} + \frac{dv}{dy} + \frac{dw}{dz}\right) = 0. \quad (886)$$

If the fluid be a homogeneous liquid we have

$$\delta = \text{a constant}, \quad \ldots \ldots \quad (887)$$

which gives the condition

$$\frac{d\delta}{dt} + \frac{d\delta}{dx}u + \frac{d\delta}{dy}v + \frac{d\delta}{dz}w = 0, \quad \ldots \quad (888)$$

310   MECHANICS OF FLUIDS.

and the equation of continuity for such liquids becomes

$$\frac{du}{dx} + \frac{dv}{dy} + \frac{dw}{dz} = 0. \quad \ldots \ldots \quad (889)$$

If the fluid be a gas, then we have by Boyle's or Charles's law a relation connecting the pressure and density, expressed generally by

$$f(p, \delta) = 0. \quad \ldots \ldots \quad (890)$$

For incompressible fluids the five necessary equations connecting u, v, w, p and δ are (882), (887) and (889); and for compressible fluids, (882), (886) and (890).

252. *Solution when the Potential and Velocity Functions are Exact Differentials.*—Let ψ be a function connected with the velocities by the equation

$$u\,dx + v\,dy + w\,dz = d\psi, \quad \ldots \ldots \quad (891)$$

and suppose that $d\psi$ is a perfect differential: then we have

$$\left. \begin{array}{lll} u = \dfrac{d\psi}{dx}, & v = \dfrac{d\psi}{dy}, & w = \dfrac{d\psi}{dz}, \\[2mm] \dfrac{du}{dx} = \dfrac{d^2\psi}{dx^2}, & \dfrac{dv}{dy} = \dfrac{d^2\psi}{dy^2}, & \dfrac{dw}{dz} = \dfrac{d^2\psi}{dz^2}. \end{array} \right\} \quad \ldots \quad (892)$$

Substituting these values in Eqs. (882), we readily obtain

$$\left. \begin{array}{l} \dfrac{1}{\delta}\dfrac{dp}{dx} = X - \dfrac{d^2\psi}{dx\,dt} - \left(\dfrac{d\psi}{dx}\dfrac{d^2\psi}{dx^2} + \dfrac{d\psi}{dy}\dfrac{d^2\psi}{dx\,dy} + \dfrac{d\psi}{dz}\dfrac{d^2\psi}{dx\,dz}\right), \\[2mm] \dfrac{1}{\delta}\dfrac{dp}{dy} = Y - \dfrac{d^2\psi}{dy\,dt} - \left(\dfrac{d\psi}{dy}\dfrac{d^2\psi}{dy^2} + \dfrac{d\psi}{dz}\dfrac{d^2\psi}{dy\,dz} + \dfrac{d\psi}{dx}\dfrac{d^2\psi}{dx\,dy}\right), \\[2mm] \dfrac{1}{\delta}\dfrac{dp}{dz} = Z - \dfrac{d^2\psi}{dz\,dt} - \left(\dfrac{d\psi}{dz}\dfrac{d^2\psi}{dz^2} + \dfrac{d\psi}{dx}\dfrac{d^2\psi}{dz\,dx} + \dfrac{d\psi}{dy}\dfrac{d^2\psi}{dy\,dz}\right). \end{array} \right\} \quad (893)$$

Multiplying these equations by dx, dy, dz, respectively, and adding, we have

$$\frac{dp}{\delta} = (X\,dx + Y\,dy + Z\,dz) - d\frac{d\psi}{dt} - \frac{1}{2}d\left(\frac{d\psi^2}{dx^2} + \frac{d\psi^2}{dy^2} + \frac{d\psi^2}{dz^2}\right), \quad (894)$$

which, if $X\,dx + Y\,dy + Z\,dz = d\,\Pi$ be also a perfect differential, may be written

$$\frac{dp}{\delta} = d\Pi - d\frac{d\psi}{dt} - \frac{1}{2}d\left(\frac{d\psi^2}{dx^2} + \frac{d\psi^2}{dy^2} + \frac{d\psi^2}{dz^2}\right). \quad \ldots \quad (895)$$

If we substitute for the differential coefficients of the velocity function their values derived from Eq. (892), and distinguish by a subscript the differential of the velocity as a function of the time from that as a function of the coordinates of position of the element, Eq. (895) can be written

$$\frac{dp}{\delta} = d\Pi - d_t(V^2) - \frac{1}{2}d_s(V^2). \quad \ldots \ldots \quad (896)$$

In this form it is evident that the equation can be integrated, (1) when δ is constant, and (2) when δ, being a variable, is such a function of p as to reduce $\frac{dp}{\delta}$ to a known integrable form. The first case applies to liquids, and the second to gases.

Flow of Perfect Liquids through Vessels.

253. Let the weight be the only force acting on the fluid, and take the axis of z vertical and positive upwards; then will

$$X = 0, \quad Y = 0, \quad \text{and} \quad Z = -g.$$

The component velocities depend only on z, and therefore the velocity function Eq. (891), considered as a function of the coordinates of the element m, may be written

$$\psi = \int V\,dz, \quad \ldots \ldots \ldots \quad (897)$$

in which V is the actual velocity at any point whose coordinate is z. Let s_0, s_1 and s be the areas of the upper, lower and any intermediate horizontal section of the vessel whose vertical coordinates are z_0, z_1, z, and let V_0, V_1 and V be the corresponding velocities of the fluid element in passing these sections at the time t. Then, since the same quantity of liquid must flow through every horizontal section in the same time, the principle of continuity requires that

$$V_0 s_0 = Vs = V_1 s_1. \quad \ldots \ldots \quad (898)$$

Substituting the value of $V = V_1 \frac{s_1}{s}$ in Eq. (897), we have

$$\psi = V_1 s_1 \int \frac{dz}{s}; \quad \ldots \ldots \quad (899)$$

whence, differentiating with respect to t, and recollecting that the velocity V_1 is negative, we have

$$\frac{d\psi}{dt} = -\frac{dV_1}{dt} s_1 \int \frac{dz}{s}. \quad \ldots \ldots \quad (900)$$

Substituting in Eq. (894), we have, after multiplying by δ and integrating,

$$p = -\delta gz + \delta s_1 \frac{dV_1}{dt} \int \frac{dz}{s} - \delta \frac{V_1^2}{2} \cdot \frac{s_1^2}{s^2} + C, \quad \ldots \quad (901)$$

a general formula, from which the pressure p may be found corresponding to any horizontal section, when the size and figure of the vessel are known. Let p_0 and p_1 be the pressures at s_0 and s_1 respectively; then Eq. (901) becomes

$$p_0 - p_1 = -\delta g(z_0 - z_1) + \delta s_1 \frac{dV_1}{dt} \int_1^0 \frac{dz}{s} - \delta \frac{V_1^2}{2} \left(\frac{s_1^2 - s_0^2}{s_0^2} \right), \quad (902)$$

from which V_1, the velocity of discharge through s_1, can be found.

FLOW OF PERFECT LIQUIDS THROUGH VESSELS. 313

254. There are two cases to consider: (1) the upper surface may be supposed kept at a constant level by a sufficient supply of the liquid, and (2) the vessel may be allowed to empty itself.

FIRST CASE. *The Upper Surface at a Constant Level.*—In this case $z_0 - z_1$ is a constant, and may be placed equal to h, which is called the *head of fluid*. Let

$$\left. \begin{array}{l} A = 2s_1 \int_0^h \dfrac{ds}{s}, \\[6pt] B = 2g\left(h + \dfrac{p_0 - p_1}{\delta g}\right), \\[6pt] C = \dfrac{s_1^2}{s_0^2} - 1. \end{array} \right\} \quad \ldots \ldots (903)$$

Substituting these symbols for their values in Eq. (902) and solving with respect to dt, we have

$$dt = \frac{A\, dV_1}{B + CV_1^2}. \quad \ldots \ldots (904)$$

Three cases may occur depending on the value of C. These are:
  (1) $C > 0$, which is the case when $s_1 > s_0$;
  (2) $C = 0$ when $s_1 = s_0$; and
  (3) $C < 0$ when $s_1 < s_0$;

First. $C > 0$.
Then Eq. (904) may be written

$$dt = \frac{A}{\sqrt{BC}} \frac{\sqrt{\dfrac{C}{B}}\, dV_1}{1 + \dfrac{C}{B}V_1^2}. \quad \ldots \ldots (905)$$

Integrating, and supposing that $V_1 = 0$ when $t = 0$, we have

$$t = \frac{A}{\sqrt{BC}} \tan^{-1} V_1 \sqrt{\frac{C}{B}};$$

whence

$$V_1 = \sqrt{\frac{B}{C}} \tan \frac{\sqrt{BC}}{A} t. \quad \ldots \quad (906)$$

This value shows that the velocity increases very rapidly with the time, and becomes infinite when

$$\frac{\sqrt{BC}}{A} t = \frac{\pi}{2},$$

or, in the finite time,

$$t = \frac{\pi A}{2\sqrt{BC}}, \quad \ldots \quad (907)$$

a result which is not possible to obtain in any actual case. The discrepancy is readily seen to arise from the impossibility of complying with the conditions of continuity of the liquid. For, since the area of the lower section is greater than the upper, the consecutive sections in falling cannot completely occupy the enlarging areas under the action of gravity alone.

Second. $C = 0$.

In this case the integration of Eq. (904) gives

$$t = \frac{A}{B} V_1;$$

whence

$$V_1 = \frac{B}{A} t = \frac{g \left(h + \frac{p_0 - p_1}{\delta g} \right)}{s_1 \int_0^h \frac{dz}{s}} t; \quad \ldots \quad (908)$$

that is, the velocity varies directly with the time as in the case of any falling body. If we neglect the difference of pressure $p_0 - p_1$ on the upper and lower surfaces, which in this case is the difference of the atmospheric pressures, the coefficient of t becomes g, and the case is the same as that of a solid falling freely in vacuo under the action of its own weight. This case can also never be an actual one.

Third. $C < 0$ or *negative*.
Then Eq. (904) may be written

$$dt = \frac{A\,dV_1}{B - CV_1^2} = \frac{A}{2\sqrt{CB}} \frac{dV_1}{\frac{1}{2}\left(\sqrt{\frac{B}{C}} - \sqrt{\frac{C}{B}}V_1\right)} \quad . \quad (909)$$

Integrating, we have

$$t = \frac{A}{2\sqrt{BC}} \log \frac{\sqrt{B} + V_1\sqrt{C}}{\sqrt{B} - V_1\sqrt{C}}; \quad . \quad . \quad . \quad (910)$$

whence

$$V_1 = \frac{e^{\frac{2\sqrt{BC}}{A}t} - 1}{e^{\frac{2\sqrt{BC}}{A}t} + 1} \sqrt{\frac{B}{C}}; \quad . \quad . \quad . \quad (911)$$

Since t enters the exponent of e, the limit of V_1 will be

$$V_1 = \sqrt{\frac{B}{C}} = \sqrt{\frac{2g\left(h + \frac{p_0 - p_1}{\delta g}\right)}{1 - \frac{s_1^2}{s_0^2}}}, \quad . \quad . \quad (912)$$

and this value of V_1 will the sooner occur as the value of C approaches -1; that is, as the area of the section of discharge is insignificant with respect to that of the upper surface. Supposing the discharge to take place in the atmosphere, then $p_0 - p_1$ may be neglected, and if the ratio $\frac{s_1}{s_0}$ be also negligible, we will have

$$V_1^2 = 2gh; \quad . \quad . \quad . \quad . \quad . \quad (913)$$

that is, the *velocity of discharge of a liquid through a small orifice, when this is due only to the action of its weight, is the same as that acquired by a body in falling over a height equal to the depth of the orifice below the upper surface.* This law was discovered experimentally by Torricelli, and is known as *Torricelli's Law*.

255. SECOND CASE. *The Vessel is allowed to Empty Itself.*—In this case, $h = z_0 - z_1$ in Eq. (902) will be variable, decreasing with time, and s_0 will also vary except where the vessel has a uniform area of cross-section. Let the velocity of discharge be given by

$$V_1 = \sqrt{2gH}, \quad \ldots \ldots \ldots (914)$$

in which H is variable; whence

$$dV_1 = \frac{g\,dH}{\sqrt{2gH}}. \quad \ldots \ldots \ldots (915)$$

By the law of continuity the same quantity of liquid must pass each section during the same time; that is, considering the variable upper and the constant lower sections, we have

$$-s_0\,dh = s_1 V_1\,dt,$$

whence

$$dt = -\frac{s_0\,dh}{s_1 \sqrt{2gH}}. \quad \ldots \ldots \ldots (916)$$

Integrating, we have

$$t = C - \frac{1}{s_1 \sqrt{2g}} \int \frac{s_0\,dh}{\sqrt{H}}. \quad \ldots \ldots (917)$$

In this equation s_0, h and H are variables; therefore, before integration can be effected, we must find in the particular problem to be solved the value of H and s_0 in terms of h. The relation between s_0 and h can be found when the figure of the vessel is known. To find the relation between H and h, we eliminate V_1, dV_1 and dt from Eq. (902) by means of Eqs. (914), (915) and (916), and the resulting equation will give H in terms of h, which, when substituted in Eq. (917), reduces the second member to a function of a single variable h.

256. Vessel of uniform cross-section discharging liquid through a small orifice.—In this case H will always be equal to the variable head h, and Eq. (917) becomes

$$t = C - \frac{2s_0}{s_1 \sqrt{2g}} \sqrt{h}. \quad \ldots \ldots \quad (918)$$

To find the value of C, suppose $h = h'$ when $t = 0$; then

$$t = \frac{2s_0}{s_1 \sqrt{2g}} (\sqrt{h'} - \sqrt{h}). \quad \ldots \ldots \quad (919)$$

The time required for the vessel to empty itself, h being then zero, is

$$t = \frac{2s_0}{s_1 \sqrt{2g}} \sqrt{h'}. \quad \ldots \ldots \quad (920)$$

Comparing this interval with the time required by the same vessel to discharge an equal volume of fluid when kept constantly full, we have in the latter case

$$V_1 = \sqrt{2gh'};$$

the volume discharged in the time t is

$$s_1 t \sqrt{2gh'};$$

which under the given conditions is equal to the volume of liquid in the vessel when full, or

$$s_1 t \sqrt{2gh'} = s_0 h',$$

whence

$$t = \frac{s_0}{s_1} \sqrt{\frac{h'}{2g}}. \quad \ldots \ldots \quad (921)$$

Therefore the time required for a vessel of uniform cross-section area to discharge itself through a small orifice is double that required to discharge an equal volume of liquid when kept full.

Steady Flow of Fluids.

257. We see, Eq. (896), that in the general case the velocity of a fluid element is a function both of the *time* and *the coordinates of position*. When the velocity at any point is independent of the time, the term $d_t(V^2)$ in Eq. (896) becomes zero, and the flow is said to be *steady*. In such cases the quantity of fluid passing any section is constant, and is equal to that passing any other section in the same time.

There are two cases to consider: (1) The density constant, as in homogeneous liquids. For this case we have, Eq. (896), taking the weight to be the only force acting and assuming z vertical and positive upward,

$$\frac{dp}{\delta} = -g\,dz - \frac{1}{2}d(V^2), \quad \ldots \ldots (922)$$

or, integrating and dividing by g,

$$\frac{p}{\omega} + z + \frac{V^2}{2g} = H, \quad \ldots \ldots (923)$$

an equation giving the relation between the pressure and velocity at any point.

To determine the meaning of the constant of integration, H, let $V = 0$. Then taking $p = 0$, which gives the free surface of the liquid, we have

$$z = H.$$

That is, *H is the height of the free surface of the liquid above the plane of reference.* It is composed of three parts: the height due to the pressure on any fluid element, the height of that element above the plane of reference, and the height due to its velocity.

To illustrate, let AB, Fig. 90, be the free surface of a liquid maintained at a constant level, ss' any stream-line of the liquid in steady flow, and XY the plane of reference; let z_0 and z_1 be the positions of the element under consideration while passing the sections a_0 and a_1. Then the sums of the three heights at the two sections are equal to each other, and each sum equal to H; the same is true for any other section. This principle is called *Bernouilli's Law*.

258. The height of the free surface above any fluid element is called the *hydrostatic head* for the position of the element, while the height to which the fluid would rise in a small tube at that position in steady flow is called the *hydrodynamic head*. Thus for

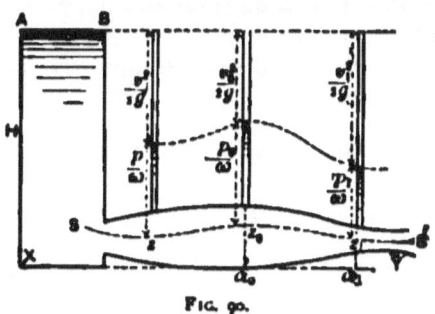

Fig. 90.

the point z_1, Fig. 90, the hydrostatic head is $H - z_1$, and the hydrodynamic head is that indicated in the piezometer tube at a_1.

Let the liquid flow through an orifice at the height z_1; then

$$\frac{p_1}{\omega} + z_1 + \frac{V_1^2}{2g} = \frac{p_0}{\omega} + z_0 + \frac{V_0^2}{2g}, \quad \dots \quad (924)$$

the first member referring to the conditions at the orifice, and the second to those at the free surface. Omitting the change in air-pressure between z_1 and z_0, which for small values of $z_0 - z_1$ is permissible, we have $p_1 = p_0$; and if the area of the free surface be so great as to make V_0 inappreciable compared with V_1, we shall have

$$V_1^2 = 2g(z_0 - z_1) = 2gh. \quad \dots \quad (925)$$

That is, the velocity of egress is that due to the *head of fluid*, the law of Torricelli.

The head of fluid in Eq. (925) may be replaced by a head due to an equivalent pressure p, as in the pump of a fire-engine. This gives us

$$V_1^2 = 2g\frac{p}{\omega},$$

p being the pressure on the piston of the pump.

259. We see, Eq. (912), that the velocity of discharge is independent of the figure of the vessel and depends only on the ratio of the areas of the upper and lower sections. If then we suppose these to be determined upon in any given case, the velocity at the orifice will be constant, the flow steady, and therefore $\frac{dV_1}{dt}$ will be zero. Eq. (902) will then give for the pressure at any point z,

$$p = p_1 - \delta g(z - z_1) + (\delta g h + p_0 - p_1)\frac{s_0^2}{s^2} \cdot \frac{s^2 - s_1^2}{s_0^2 - s_1^2}. \quad (926)$$

Hence *in uniform flow through a vessel the pressure on any horizontal stratum is independent of the figure of the vessel, and depends only on its distance $z - z_1$ above the orifice, the area s of the stratum, and the ratio of the upper and lower areas.* It also appears that if any section be equal in area to that of the discharge, the pressure p at that area will be less than that at the level of the discharge by that due to a head of fluid equal to $z - z_1$. In this case a partial vacuum will exist, and if a pipe be connected with a lower reservoir to this point, water would be drawn from it and discharged through the vessel, as indicated in Fig. 91.

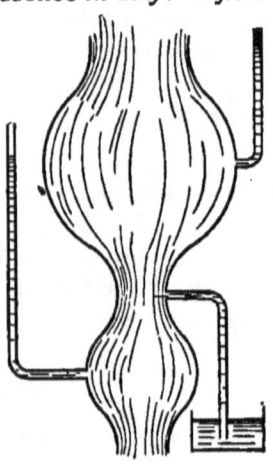

Fig. 91.

STEADY FLOW OF FLUIDS.

260. (2) *Steady Flow of Gases.*—For gases we have, Eq. (896),

$$\frac{dp}{\delta} = -gdz - \frac{1}{2}d(V^2). \quad \ldots \ldots (927)$$

Dividing by g, we have

$$\frac{dp}{\delta g} = -dz - \frac{1}{2g}d(V^2), \quad \ldots \ldots (928)$$

or, since $\dfrac{1}{\delta g} = v =$ the specific volume,

$$vdp = -dz - \frac{1}{2g}d(V^2). \quad \ldots \ldots (929)$$

From Eq. (823) we have

$$pv = R\tau,$$

or

$$vdp = R\tau\frac{dp}{p}; \quad \ldots \ldots (930)$$

and substituting in Eq. (929),

$$R\tau\frac{dp}{p} = -dz - \frac{1}{2g}d(V^2). \quad \ldots \ldots (931)$$

The integration of this equation will give the conditions of flow.

(1) Let the vessel through which steady flow takes place be impervious to heat; then $dQ = 0$, and we shall have

$$R\tau\frac{dp}{p} = C_p d\tau, \quad \ldots \ldots (932)$$

and Eq. (931) becomes

$$C_p d\tau = -dz - \frac{1}{2g}d(V^2). \quad \ldots \ldots (933)$$

Integrating between limits, we have

$$\frac{1}{2g}(V_1^2 - V_0^2) = z_0 - z_1 + C_p(\tau_0 - \tau_1). \quad . \quad . \quad (934)$$

From this equation V_1, the velocity of egress, may be determined when τ_1 is known. The conditions at the point z_0 are supposed known.

To reduce this equation to a form free from τ_1, we have, from Poisson's law,

$$\frac{v_1}{v_0} = \left(\frac{p_0}{p_1}\right)^{\frac{1}{k}}$$

and

$$\frac{p_1 v_1}{p_0 v_0} = \frac{R\tau_1}{R\tau_0} = \frac{\tau_1}{\tau_0}.$$

Substituting the value of $\frac{v_1}{v_0}$, Eq. (840), in the last equation, we have

$$\frac{\tau_1}{\tau_0} = \frac{p_1}{p_0} \times \left(\frac{p_0}{p_1}\right)^{\frac{1}{k}} = \left(\frac{p_1}{p_0}\right)^{\frac{k-1}{k}};$$

whence

$$\tau_1 = \tau_0 \left(\frac{p_1}{p_0}\right)^{\frac{k-1}{k}};$$

which substituted in Eq. (934) gives

$$\frac{V_1^2 - V_0^2}{2g} = z_0 - z_1 + C_p \tau_0 \left[1 - \left(\frac{p_1}{p_0}\right)^{\frac{k-1}{k}}\right]; \quad . \quad . \quad (935)$$

from which the velocity of egress can be found when p_1 is known.

(2) If the quantity of heat be supposed to change in such a manner as to make the temperature constant, then from Eq. (931) we shall have, by integrating between limits,

$$\frac{1}{2g}(V_1^2 - V_0^2) = z_0 - z_1 + R\tau.\text{Nap. log}\frac{p_0}{p_1}. \quad . \quad . \quad (936)$$

This is called the *Theorem of Navier*. But as air is a poor conductor of heat, this formula is applicable only when the variation of temperature is negligible.

If, in addition to the conditions imposed above, the air be at rest, Navier's formula reduces to

$$z_1 - z_0 = RT \text{ Nap. log} \frac{p_0}{p_1}, \quad \ldots \quad (937)$$

which is known as the *Barometric* formula, and which will be discussed later.

Equilibrium of Liquids in Motion.

261. If the acceleration terms in Eqs. (880) be zero, we shall have

$$\left. \begin{array}{l} \dfrac{1}{\delta} \dfrac{dp}{dx} = X, \\[4pt] \dfrac{1}{\delta} \dfrac{dp}{dy} = Y, \\[4pt] \dfrac{1}{\delta} \dfrac{dp}{dz} = Z; \end{array} \right\} \quad \ldots \ldots \quad (938)$$

which, when multiplied respectively by dx, dy, dz, and added, give, after reduction,

$$dp = \delta (X\,dx + Y\,dy + Z\,dz). \quad \ldots \quad (939)$$

Eqs. (938) express the condition that the intensity of the pressure is equal to that of the resultant of the applied forces.

If the uniform motion be rotation about an axis, a component of the resultant of the forces directly applied must be considered in deflecting the molecule from its right-line path. Hence

$\sqrt{\overline{mX'}^2 + \overline{mY'}^2 + \overline{mZ'}^2}$ is one component of the resultant of the forces directly applied, the other component being the centrifugal force.

mX, mY, mZ, represent the components of the resultant of the forces directly applied: then $mX = mX' +$ the x component of the centrifugal force. Let ω be the angular velocity, ρ the radius of curvature, and z the axis of rotation: then we shall have for the component acceleration toward the axis

$$\rho\omega^2 \frac{x}{\rho} = \omega^2 x \quad \text{and} \quad \rho\omega^2 \frac{y}{\rho} = \omega^2 y,$$

and hence

$$mX = mX' + m\omega^2 x,$$

or

$$X = X' + \omega^2 x;$$

and similarly,

$$Y = Y' + \omega^2 y.$$

Substituting in Eq. (939), we have

$$dp = \delta\left[(X' + \omega^2 x)\,dx + (Y' + \omega^2 y)\,dy + Z'\,dz\right].. \quad (940)$$

For the free surface or any surface of uniform pressure, since then $dp = 0$, we shall have the condition

$$(X' + \omega^2 x)\,dx + (Y' + \omega^2 y)\,dy + Z'\,dz = 0, \quad .. \quad (941)$$

from which the equation of the free surface and other surfaces of uniform pressure can be found.

Ex. 1. *Liquid Rotating About a Vertical Axis.*—Let the liquid be contained in a cylindrical vessel, Fig. 92, whose axis is vertical and is the axis of rotation, and let gravity be the only force acting. Then we have

$$X' = 0, \quad Y' = 0, \quad Z' = -g,$$

and Eq. (941) becomes

$$gdz = \omega^2 (xdx + ydy).$$

Taking ω to be constant and integrating, we have

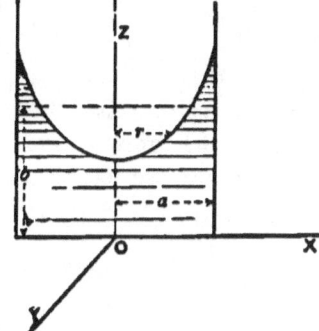

FIG. 92.

$$z = \frac{\omega^2}{2g}(x^2 + y^2) + C = \frac{\omega^2 r^2}{2g} + C, \quad \ldots \quad (942)$$

in which r is the variable distance of the free surface from the axis of rotation. Hence, under the given conditions, the free surface of a rotating liquid is that of a paraboloid of revolution.

To find the constant C, let a be the radius of the cylindrical vessel, and b the height of the liquid when at rest; then, the volume of liquid when at rest and when in rotation being the same, we have

$$\pi a^2 b = \int_0^a 2\pi r z dr. \quad \ldots \quad \ldots \quad (943)$$

If h be the height due to the velocity of the element on the cylindrical surface, we have

$$a^2 \omega^2 = 2gh.$$

Substituting the value of ω^2 in Eq. (942), we have

$$z = \frac{hr^2}{a^2} + C; \quad \ldots \quad \ldots \quad (944)$$

from which we have

$$rdr = \frac{a^2 dz}{2h}.$$

Substituting this value of rdr in Eq. (943), and integrating between the limits C and $C + h$, corresponding to the limits $r = 0$ and $r = a$, we have for the value of C,

$$C = b - \tfrac{1}{2}h.$$

Substituting this in Eq. (944), we have

$$z = \frac{hr^2}{a^2} + b - \frac{1}{2}h. \quad \ldots \ldots \quad (945)$$

We see from this that the element on the axis is depressed during rotation a distance equal to one half the height h, and the element on the surface of the cylinder is elevated an equal distance above the original level.

EX. 2. *To find the free surface of a rotating liquid whose elements are attracted towards their centre of mass by a force varying inversely as the square of the distance.*—This problem is the same as that of finding the form of the level surface on the earth's surface. Let r be the distance of any element from the centre of mass, and μ the intensity of the attraction for the unit mass at the distance unity; let α, β, γ be the angles which the action-line of the attraction makes with the axes. Then

$$X' = -\frac{\mu}{r^2}\cos\alpha = -\frac{\mu x}{r^3}, \quad Y' = -\frac{\mu y}{r^3}, \quad Z' = -\frac{\mu z}{r^3}, \quad (946)$$

which, if z be the axis of rotation, being substituted in Eq. (941) give

$$\frac{\mu}{r^3}(xdx + ydy + zdz) - \omega^2(xdx + ydy) = 0, \quad . \quad . \quad (947)$$

or

$$\frac{\mu dr}{r^2} - \frac{\omega^2}{2} d(x^2 + y^2) = 0; \quad \ldots \quad (948)$$

whence

$$\frac{\mu}{r} + \frac{\omega^2}{2}(x^2 + y^2) = C. \quad \ldots \quad (949)$$

If θ be the angle which r makes with the plane xy, we have

$$\frac{\mu}{r} + \frac{\omega^2}{2} r^2 \cos^2 \theta = C. \quad \ldots \quad (950)$$

Let the distance from the origin to the point in which the axis of z cuts the free surface be unity: then, since θ is 90° for this value of r, we have

$$C = \mu.$$

Whence Eq. (950) becomes

$$\frac{\omega^2}{2} r^2 \cos^2 \theta = (r-1)\mu, \quad \ldots \quad (951)$$

or

$$\frac{1}{2} \omega^2 \cos^2 \theta = \mu \frac{r-1}{r^2}, \quad \ldots \quad (952)$$

for the polar equation of the free surface.

The approximate solution applicable to the earth's free liquid surface may thus be found. Let $r = 1 + u$, u being small with respect to unity, as its maximum value is less than $\frac{1}{300}$: then

$$\frac{r-1}{r^2} = \frac{u}{(1+u)^2} = u - 3u^2, \text{ nearly.}$$

Substituting in Eq. (952), we have

$$u - 3u^2 = \frac{\omega^2}{2\mu} \cos^2 \theta,$$

328   MECHANICS OF FLUIDS.

whence

$$u = r - 1 = \frac{\omega^2}{2\mu}\cos^2\theta + \frac{3}{4}\frac{\omega^4}{\mu^2}\cos^4\theta,$$

and

$$r = 1 + \frac{\omega^2}{2\mu}\cos^2\theta + \frac{3}{4}\frac{\omega^4}{\mu^2}\cos^4\theta + \text{etc.}, \quad . \quad . \quad (953)$$

which is aproximately the equation of an ellipse whose eccentricity is $\sqrt{\frac{\omega^2}{\mu}}$. Therefore the free surface is very nearly that of an oblate spheroid.

Hydraulics.

262. The circumstances attending the motion of water flowing through pipes and canals, and issuing from apertures in the sides of vessels, differ considerably from the results deduced theoretically. These differences are obviously due to certain resistances which modify the supposed condition of perfect freedom of motion of the elements of actual liquids. Numerous experiments are necessary to ascertain the effect of these resisting forces and to determine the empirical laws governing their variation. The limits of the course prevent anything more than a reference to the simpler illustrations, and that which follows is a brief condensation of the conclusions reached by many experimenters.*

In the theory of the steady flow of liquids, the stream-lines have been supposed to consist of consecutive filaments moving with parallel relative motions at each normal section, and whose velocities are continuous functions of the coordinates of position of the fluid element; while, actually, the stream is more or less disturbed from time to time by detached portions of these fila-

---

* See Weisbach's Mechanics of Fluids; Encyclopedia Britannica, 9th ed., Vol. XII., art. Hydromechanics, etc.

ments, which form eddies that constantly change their extent and position and traverse the liquid in all directions. The total resistance caused by these irregular and inconstant motions varies from time to time, and its average intensity is estimated from the effects produced during periods in which the variations are negligible. Effects produced by resistances of this nature are found to be the most important.

The *viscosity* of the liquid also modifies the theoretical results, but, in the case of water, the resulting resistance is of much less importance than that arising from eddies and irregularities in the stream-lines. The resistance *due to viscosity* is assumed to vary *directly* with the area of the stratum, and with its relative velocity with respect to the adjacent stratum, and *inversely* as the thickness of the stratum; hence if a, V and T represent these quantities, and μ be the coefficient of viscosity, we have for the measure of the intensity of this resistance for an indefinitely thin layer,

$$R = \mu a \frac{dV}{dT}. \quad \ldots \quad \ldots \quad (954)$$

In water μ has been found to diminish rapidly as its temperature increases; its value for water at 77° F. according to Helmholtz is $\frac{191}{10^4}$ pounds per square foot, when the relative velocity is given in feet per second and T is expressed in feet. For air at θ° F. μ in pounds according to Maxwell is $\frac{256}{10^{14}}(461° + \theta)$; hence it is independent of the pressure and varies directly with the absolute temperature.

263. *Experimental Coefficients in the Flow of Water.*—When water issues from a vessel through a small orifice, the stream is called a *jet;* and if the jet be vertical and be caused only by the pressure arising from the weight of the water in the vessel, it will be found to rise very nearly to the height of the water-level. The small difference between the altitude of the jet and the upper surface is due to air resistance, friction and viscosity. Neglecting this difference, the velocity of discharge, called the *theoretical discharge*, is given by the law of Torricelli,

$$v = \sqrt{2gh};$$

h being the altitude of the free surface above the orifice. If α be the area of the orifice, and Q the quantity theoretically discharged in the unit time, called the *expense*, we have

$$Q = \alpha \sqrt{2gh}. \quad \ldots \ldots \ldots \quad (955)$$

Numerous experiments have been made with different heads and well-formed simple orifices, and the actual velocity in all cases is found to be very nearly 0.97 that of the theoretical. This is called the *coefficient of velocity;* and representing it by C_v, we have

$$C_v = 0.97.$$

When water flows from an orifice, the stream-lines are found to converge towards the orifice from all points in the upper surface. The motions of the liquid elements at the orifice depend on the size, position and character of the orifice; and where the orifice is in thin walls, the convergence of the filaments is found to continue for a distance beyond the orifice equal to about one half its diameter. At this point the jet has a minimum cross-section, and the property of jets to take this form is called that of *veinal contraction*. At the contracted section the stream-lines of the liquid elements are assumed to pass it in normal directions, and therefore the velocity of discharge through this section can be determined experimentally when its area and the quantity discharged in a given time are known. Both of these have been measured with considerable accuracy, and the *coefficient of contraction* C_c for sharp-edged orifices in plane surfaces has been found to be 0.64; hence the cross-section area of the jet slightly exceeds five eighths that of the orifice. If C represent the *coefficient of discharge*, its value is

$$C = C_v \times C_c = 0.97 \times 0.64 = 0.62;$$

whence the formula to be used for estimating the discharge of water from sharp-edged orifices in plane sides, whose area is α, is

$$Q_a = C\alpha \sqrt{2gh} = 0.62\alpha \sqrt{2gh}; \quad \ldots \quad (956)$$

Q_a being the quantity actually discharged. If the orifice be in a thick plane side, such that the thickness of the walls is equal to or greater than the diameter of the jet, the water adheres to the sides of the aperture and the coefficient C_c becomes equal to unity; that is, there is no contraction. This condition is also true for bell-mouthed orifices whose form is that of the contracted vein.

264. *Mouthpiece of Borda*.—This consists (Fig. 93) of a reentrant cylindrical tube $ABCD$ of the greatest length that will permit the liquid to flow from the vessel without touching the side of the vessel. When the flow has become steady, the stream-lines from the level HH' to the section $\alpha\alpha$ of greatest contraction are such as to allow but little, if any, motion to the particles near AC and BD; therefore the pressures at these points are those of the corresponding hydrostatic heads. The horizontal pressures on HE and $H'C$, being equal and opposite, counterbalance each other; so also do those on opposite sides of the vessel below FD. If s be the area of EF, and p_a the atmospheric pressure per unit of area, the horizontal pressure on EF is $(\delta gh + p_a)s$. The pressure on the surface and section $A\alpha\alpha B$ is $p_a s$; hence the resultant pressure producing momentum in the horizontal section through $\alpha\alpha$ is δghs.

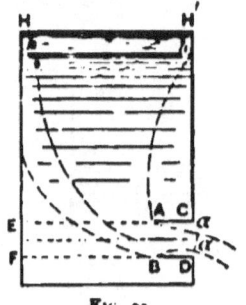

Fig. 93.

Let this pressure act long enough on a section of the fluid to give it its velocity of discharge through the contracted section, the time being represented by t, and the mass of fluid discharged in the time t by m. Let α represent the area of the contracted section. Then

$$\delta ghst = mV = \delta\alpha V^2 t,$$

or

$$\frac{\alpha}{s} = \frac{gh}{V^2};$$

whence we have, by the application of Torricelli's law,

$$\frac{\alpha}{s} = \frac{1}{2}.$$

Therefore theoretically the area of the contracted section is one half the area of this mouthpiece, a result amply confirmed by experiment.

265. *Cylindrical Mouthpiece.*—When water flows through an external cylindrical mouthpiece whose length exceeds one and a half times the shortest dimension of the orifice, it is found that the vein after leaving the contracted section opens out and fills the cylinder completely, leaving it without further contraction. Let Fig. 94 represent a vessel discharging water from a constant head through such a mouthpiece. Let the horizontal XX' at the distance h below the upper surface pass through the centre of the orifice and be the line of reference. Let s, v, p, be the area, velocity and pressure, respectively, at the contracted section, and s_1, v_1, p_1, be the corresponding values

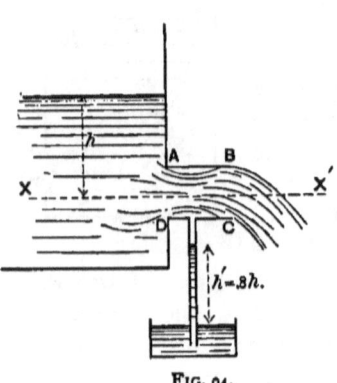

Fig. 94.

at the section BC. After passing the contracted section, the water experiences a somewhat abrupt change of velocity, due to the formation of eddies in the vacant space around this section, and there is consequently what is called a *loss of head due to shock*, which always occurs at sudden enlargements of the cross-section of flowing water. A certain amount of energy is dissipated from this cause, whose value can be estimated as follows:

Let Q be the quantity of water that passes any section in the time dt, and consider the two sections AD and BC: then the pressure parallel to XX' is

$$ps + p_0(s_1 - s) - p_1 s_1,$$

p_0 being the pressure on the annular space surrounding the contracted section.

The impulsion due to this pressure is

$$(ps + p_0(s_1 - s) - p_1 s_1)dt;$$

and the change of momentum in the horizontal direction is

$$\delta Q(v_1 - v) = \delta s_1 v_1 (v_1 - v)dt.$$

Placing the change of momentum equal to the intensity of the impulsion which causes it, and supposing $p_0 = p$, which experiment shows is nearly true, we have after reduction

$$p - p_1 = \delta v_1 (v_1 - v). \quad \ldots \ldots \quad (957)$$

Dividing by the specific gravity ω, we have

$$\frac{p}{\omega} - \frac{p_1}{\omega} = \frac{v_1(v_1 - v)}{g};$$

whence, by adding to both members $\frac{v^2}{2g}$, we have

$$\frac{p}{\omega} + \frac{v^2}{2g} = \frac{p_1}{\omega} + \frac{v_1^2}{2g} + \frac{(v - v_1)^2}{2g}. \quad \ldots \quad (958)$$

We see from this that the energy lost in shock is $\dfrac{(v-v_1)^2}{2g}$ foot-pounds per pound of liquid.

In the case under consideration we have for the head at any point where the velocity is approximately zero,

$$h + \frac{p'}{\omega},$$

in which p' is the atmospheric pressure. Placing this equal to the head at the sections AD and BC, we have

$$h + \frac{p'}{\omega} = \frac{p}{\omega} + \frac{v^2}{2g} = \frac{p_1}{\omega} + \frac{v_1^2}{2g} + \frac{(v-v_1)^2}{2g}; \quad \ldots \quad (959)$$

from which, if the flow be into the atmosphere and therefore $p' = p_1$, we have

$$h = \frac{v_1^2 + (v - v_1)^2}{2g}.$$

Substituting for v its value $\dfrac{v_1}{C_c}$, C_c being the coefficient of contraction 0.64, we have

$$h = \frac{v_1^2}{2g}\left[1 + \left(\frac{1}{C_c} - 1\right)^2\right] = \frac{v_1^2}{2g} 1.314. \quad \ldots \quad (960)$$

Solving with reference to v_1, we have

$$v_1 = 0.87\sqrt{2gh},$$

which, since the coefficient of velocity is found by experiment to be 0.82, differs no more than might be expected when the viscosity and friction are not taken into account.

To show that the pressure p at the contracted section is much less than the atmospheric pressure, we have, from Eq. (959),

$$\frac{p_1-p}{\omega} = \frac{v^2-v_1^2-v^2+2vv_1-v_1^2}{2g} = \frac{2vv_1-2v_1^2}{2g} = 2v_1^2 \cdot \frac{\left(\frac{v}{v_1}-1\right)}{2g}.$$

Substituting the value of v_1^2, Eq. (960), and $\frac{1}{0.64}$ for $\frac{v}{v_1}$ and reducing, we have

$$\frac{p_1-p}{\omega} = 0.8h;$$

whence

$$p = p_1 - 0.8\omega h \text{ per sq. ft.}$$

Therefore, if a pipe be connected with a reservoir at a lower level and open into the annular space about the contracted vein in the discharge-pipe, the atmospheric pressure will force the water from the reservoir up to a height of $0.8h$.

266. *Divergent Mouthpiece*.—In some well-formed divergent mouthpieces, as in Fig. 95, the flow is continuous and the velocity of egress nearly that due to the head of fluid. In such cases the velocity at the contracted section CD is greater than $\sqrt{2gh}$, and the pressure at that section is therefore less than the atmospheric pressure. However, continuity of flow ceases before the pressure reaches zero, and the limit of pressure for continuity of flow must be determined by experiment.

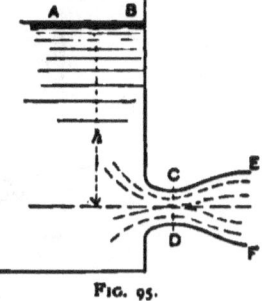

Fig. 95.

Assuming steady flow and applying Eq. (926), we have for the pressure at the contracted section

$$p = p_1 + \omega h\left(1 - \frac{s_1^2}{s^2}\right). \quad \ldots \quad (961)$$

If the contracted section be about 0.7 that of the orifice, we shall have

$$p = p_1 - \omega h.$$

That is, in a pipe leading from the contracted section to a reservoir below, the water would rise to a height equal to the head of fluid. If the height from the lower reservoir to the contracted section be not too great, the water will rise to this section and flow out through the mouthpiece. This is the principle of what is called the *Jet Pump*. It applies only within the limits of continuity of flow, and when the velocity of egress and the contraction are known the height through which the water may be raised is readily computed.

When the constant head of water and the character and form of the orifice are known, the experimental coefficients of contraction and velocity can be applied to the theoretical value of the discharge, and we can obtain, as in the above examples, practical formulas for the actual discharge, which are of use in gauging water or other liquids.

267. *The Hydraulic Gradient.*—Let steady flow take place

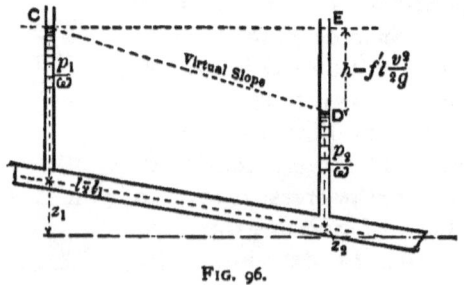

Fig. 96.

through a straight pipe of uniform cross-section, as in Fig. 96. Then we have, by Bernouilli's law,

$$\frac{p}{\omega} + z + \frac{v^2}{2g} = \frac{p'}{\omega} + z' + \frac{v^2}{2g},$$

or
$$\frac{p}{\omega} + z = \frac{p'}{\omega} + z'.$$

That is, the water rests at the same level in the piezometers C and E, and the work done by gravity in moving the liquid from z to z' is stored up in the increase of head $z - z'$. This supposes the fluid to be perfect and without friction. However, in actual cases the friction between the liquid and the pipe works against gravity, and the head at E is decreased by an amount depending on the quantity of work of friction.

Assuming a pipe of circular cross-section of radius r, we have for the elementary quantity of work done by gravity in the unit of time,

$$\omega \pi r^2 v dz;$$

for the corresponding work done by the pressure,

$$\pi r^2 v dp;$$

and for the elementary work of friction,

$$F \, 2\pi r v dl,$$

F being the frictional resistance per unit of area of contact between the water and pipe. Hence we have for uniform flow

$$\omega \pi r^2 v dz + \pi r^2 v dp + F \, 2\pi r v dl = 0. \quad \ldots \quad (962)$$

F varies as the square of the velocity; and writing

$$F = f v^2,$$

we obtain from Eq. (962)

$$dz + \frac{dp}{\omega} + \left(f \frac{4g}{\omega r}\right) \frac{v^2}{2g} dl = 0. \quad \ldots \quad (963)$$

Integrating and writing

$$f\frac{4g}{\omega r} = f',$$

we have

$$z + \frac{p}{\omega} + f'l\frac{v^2}{2g} = \text{a constant} = H - \frac{v^2}{2g}, \quad . . \quad (964)$$

H being the total head. The head lost by friction is therefore $f'l\frac{v^2}{2g}$, which varies directly as the length of the pipe and the height due to the velocity of flow. For a particular case the latter is constant, and finding the head lost in any distance we obtain the right line CD; this gives, by the vertical distance between it and the pipe, the head due to pressure at any point. Thus at D the head due to pressure is Dz_1, and the head lost by friction is ED. The line CD is called the *Hydraulic Gradient* or *Line of Virtual Slope*. If the pipe be laid parallel to this line the pressure will be constant, and we shall have

$$z + f'l\frac{v^2}{2g} = \text{a constant}. \quad . \quad . \quad . \quad . \quad (965)$$

The slight change in the length l is not considered, and the velocity of flow is supposed to be the same as before.

Applications.

268. Hydraulic Machines are divided into two classes: (1) those designed to utilize the energy of a head of water in doing useful work, such as *water-wheels, water-pressure engines* and *turbines*, and (2) those designed to lift water from a lower to a higher level, such as *pumps* and the *hydraulic ram*. The essential parts of every contrivance used to employ the energy of a natural fall of water are:

(1) A supply channel leading the water from the highest accessible level to the site of the machine.

(2) A tail-race or discharge-pipe delivering the water after having done its work at the lowest convenient level.

(3) A waste-weir to permit the escape of surplus water.

(4) The motor with its regulating machinery.

The energy available for any hydraulic motor for a total head of H feet is

$$(H - h_r - h_w) Q\omega \text{ foot-pounds,}$$

in which h_r is the loss of head due to the aggregate resistances at the head and tail races, and h_w is the head lost by waste or leakage. Of this amount of available energy only a portion can be employed in doing useful work, the remainder being used up in overcoming the wasteful resistances of friction, etc., as described in Art. 201 for machines in general.

Water Motors.

269. *Water-wheels.*—The ordinary water-wheels are classified as *overshot*, *high-breast* and *undershot* wheels. In the first, Fig. 97,

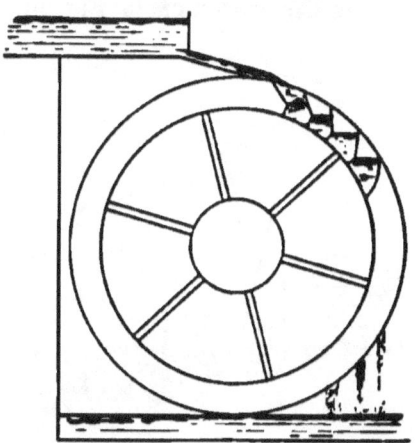

FIG. 97.

the water is projected into the buckets at the summit of the wheel, and acts first by impulse and then by its weight. In the

high-breast wheel, Fig. 98, the water is received in the buckets on the side toward the water supply, and it is prevented from

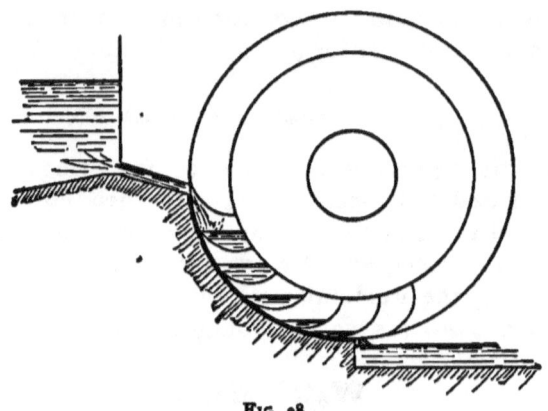

FIG. 98.

flowing out of the buckets freely by a curved wall. The undershot wheel receives the water as in Fig. 99, which shows the

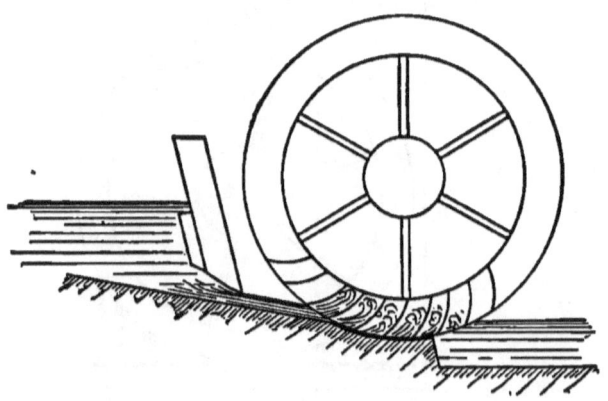

FIG. 99.

form of the *Poncelet* wheel, and for the greatest efficiency the water should fall vertically from the buckets when it leaves the wheel. In all cases a sliding sluice-gate regulates the supply according to the head of fluid.

270. Turbines.—A water-wheel (Fig. 100) to the vanes of which the water is conducted by curved guide-plates is called a

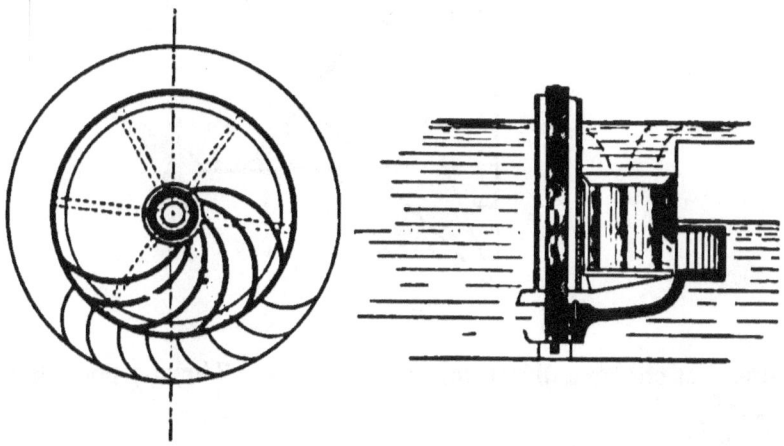

FIG. 100.

turbine. The wheel usually revolves in a horizontal plane. There are four general classes:

1. *Outward-flow turbines,* in which the water flows horizontally outward between the vanes.

2. *Inward-flow turbines,* in which the water flows horizontally inward between the vanes.

3. *Parallel-flow turbines,* in which the water flows between the vanes in the direction of the axis of the wheel.

4. *Mixed-flow turbines,* consisting of a combination of the parallel flow with either the outward or inward flow.

When the wheel revolves in a vertical plane similar principles determine their classification.

271. *Theory of Turbines.*—When a fluid changes the direction of its motion the centrifugal force at any point is $dM\dfrac{v^2}{\rho}$. Let the flow be steady through a pipe of uniform cross-section, and let it be required to find the resistance developed in any direction. Take the axis of y to be the radius of curvature where the

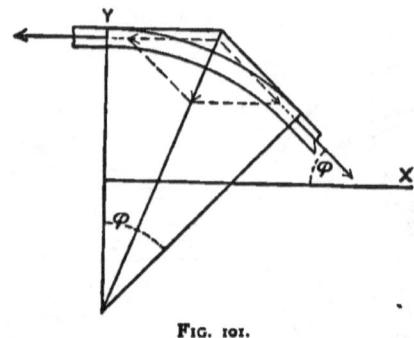

FIG. 101.

pipe first changes direction, as in Fig. 101. The components of $dM\dfrac{v^2}{\rho}$ are

$$\left.\begin{aligned} dX &= dM\frac{v^2}{\rho}\sin\phi, \\ dY &= dM\frac{v^2}{\rho}\cos\phi. \end{aligned}\right\} \quad \cdots \cdots (966)$$

Let α be the cross-section of the stream, and we have

$$dM = \delta\alpha\rho d\phi,$$

and Eqs. (966) reduce to

$$\left.\begin{aligned} dX &= \delta\alpha v^2 \sin\phi d\phi, \\ dY &= \delta\alpha v^2 \cos\phi d\phi. \end{aligned}\right\} \quad \cdots \cdots (967)$$

Integrating between the limits o and ϕ, we have

$$\left.\begin{aligned} X &= \delta\alpha v^2 (1 - \cos\phi), \\ Y &= \delta\alpha v^2 \sin\phi. \end{aligned}\right\} \quad \cdots \cdots (968)$$

Let M be the mass of fluid passing any section in a unit of time, and we have

$$M = \delta a v,$$

and substituting in Eqs. (968),

$$\left. \begin{array}{l} X = Mv(1 - \cos \phi), \\ Y = Mv \sin \phi. \end{array} \right\} \quad \cdots \quad (969)$$

For a change of direction of 90°, 180° and 360°, we have

$$\left. \begin{array}{ll} (90°) & X = Y = Mv, \\ (180°) & X = 2Mv, \ Y = 0, \\ (360°) & X = Y = 0. \end{array} \right\} \quad \cdots \quad (970)$$

These results might have been assumed directly from Art. 23, since the force required to generate in M the velocity v *in a unit of time* is Mv, and the force required to check this momentum in a given direction *in the same time* is also Mv. In the general case, Eqs. (969), the change of momentum in the direction of X is $Mv \cos \phi$, and the remaining momentum in that direction is therefore $Mv(1 - \cos \phi)$. The change of momentum in the direction of Y is $Mv \sin \phi$, and since the whole momentum was before in the direction of X, the resistance in the direction of Y is $Mv \sin \phi$.

The resultant of X and Y is

$$R = \sqrt{X^2 + Y^2} = Mv \sqrt{2(1 - \cos \phi)}; \quad (971)$$

and if forces be applied in the direction of the tube to produce the same resultant, we shall have

$$T^2 + T^2 - 2T^2 \cos \phi = 2M^2v^2(1 - \cos \phi), \quad (972)$$

or

$$T = Mv. \quad \cdots \quad \cdots \quad (973)$$

That is, the tension due to the flow is independent of the change of direction, being constant throughout.

If a tube be perfectly flexible and fixed at the ends the fluid pressure cannot change the form of the tube, and it will still be perfectly flexible while a perfect fluid is flowing through it. (Art. 220.)

Let Mv_1 be the tangential component of the momentum of the water on entering the vanes, and Mv_2 that on leaving, the corresponding radii being r_1 and r_2. The radial components of the momentum having no moment with respect to the axis of the wheel, the moment of the momentum on entering is Mv_1r_1, and on leaving Mv_2r_2; hence we have for the effective moment $M(v_1r_1 - v_2r_2)$, and for the work done per second $M(v_1r_1 - v_2r_2)\omega$, ω being the angular velocity of the wheel. The component velocities v_1 and v_2 are of course relative to the velocity of the wheel.

In any case the efficiency of the wheel is measured by the percentage of the actual head of fluid which is utilized in *effective work*. The efficiency is evidently a maximum when $v_2 = 0$.

The values of v_1 and r_2 are determined from the construction of the wheel and the conditions of motion.

Pumps.

272. *The Sucking Pump*.—The sucking pump, Fig. 102, consists of a pipe P dipping below the water in the reservoir, a cylinder C in which a close-fitting piston moves, and two valves, the sleeping valve V and the piston valve V', both opening upwards. The *play* of the piston is the distance measured on the axis of the cylinder through which the piston moves. To explain the action of this pump, let the piston be supposed to be at the lower limit of its play, and the water in the pipe to be at the level of the water in the reservoir, both valves being closed by their own weight. As the piston is raised, the air in the cylinder below the piston expands and becomes less dense than that in the sucking pipe; and when the difference of air-pressure above and below the sleeping valve is sufficient to lift the valve, the air in the pipe will occupy both pipe and cylinder. Its elastic force will then be less than that of the external air pressing on the surface of the reservoir. Hence the water will rise in the pipe until equilibrium is established between the external atmospheric pressure and the combined air and liquid pressure in the pipe. This will occur when the water rises to a height h', such that

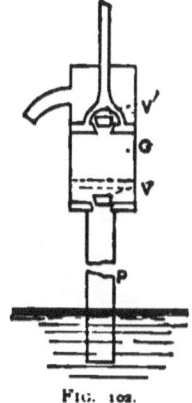

Fig. 103.

$$h' + \frac{p'}{\omega} = \frac{p_a}{\omega} \quad \text{or} \quad h' = \frac{p_a - p'}{\omega}; \quad (974)$$

in which p_a and p' are the pressures of the external and internal air, respectively, when the piston has reached the upper limit of its play. On the return stroke of the piston, the sleeping valve being closed the air in the cylinder will be condensed until its elastic force is sufficient to open the piston valve against the atmospheric pressure; then part of the air in the cylinder will escape outward. During the second upward stroke the water will rise still higher in the pipe, and at some succeeding stroke will pass through the sleeping valve into the cylinder. On the following downward stroke the water will be forced through the piston valve, and will finally be lifted to the discharge-pipe of the pump.

273. In order that the pump may work, it is evident that the play of the piston must at least be such that the elastic force of the air in the cylinder, at the end of every downward stroke, will be sufficient to open the piston valve; or, in other words, must be greater than that of the external air, increased by the weight of the piston valve. To find p, the play of the piston, let ι, Fig. 103, be the greatest height it attains above the surface of the reservoir, and x the distance from the same level to the highest plane reached by the water in a pump whose play is insufficient; and let C be the area of the piston. Then when the piston is at the upper and lower limits of its play the volumes of air in the cylinder are respectively

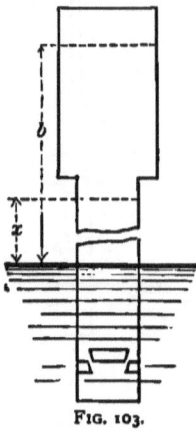

Fig. 103.

$$C(b-x) \quad \text{and} \quad C(b-x-p).$$

The elastic force of the air in the latter case will be that of the external air, or p_a. Let p' be the elastic force at the upper limit, and we have by Boyle's law

$$p'C(b-x) = p_a C(b-x-p); \quad (975)$$

whence

$$p' = \frac{b-x-p}{b-x} p_a; \quad \ldots \ldots (976)$$

or, in terms of the heights of the corresponding water columns which these pressures can support in vacuo,

$$h' = \frac{b-x-p}{b-x} h. \quad \ldots \ldots (977)$$

If x be added to h', the sum will be equal to h, since the external air-pressure is in equilibrium with the pressure within the pump, arising from the weight of the column of water x, and the elastic force of the air in the cylinder when the piston is at the upper limit of its play.

Therefore we have

$$\frac{b-x-p}{b-x} h + x = h; \quad \ldots \ldots (978)$$

whence

$$x = \frac{b \pm \sqrt{b^2 - 4ph}}{2}, \quad \ldots \ldots \quad (979)$$

When

$$p = \frac{b^2}{4h}, \text{ or } p < \frac{b^2}{4h}, \quad \ldots \ldots \quad (980)$$

x has real values, and there will be a height beyond which the water will not rise, and the pump will not work. When $p > \frac{b^2}{4h}$ the value of x is imaginary, and the play of the piston is sufficient to cause the water to be delivered at the discharge-pipe. Therefore *the play of the piston must be greater than the square of the height of the upper limit of the piston above the surface of the reservoir, divided by four times the height of the water barometer at the place.*

This is of no practical importance in the construction of pumps. The discussion assumes that the sleeping valve is not at a greater height above the water than x; but practically the piston reaches the sleeping valve at the lower limit of the stroke, and hence the play is determined without reference to the above principle.

274. *The Lifting Pump.*—Fig. 104. When the cylinder and sleeping valve are placed at the bottom of the pipe, the pump is called a *lifting pump*. It does not differ in principle, therefore, from the sucking pump, but it would be used when the mouth of the delivery pipe is at such an altitude above the reservoir as to prevent the use of the sucking pump.

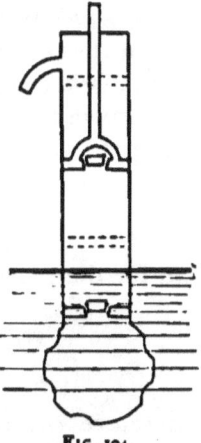

FIG. 104.

275. *The Force-pump.*—By attaching a lateral pipe to the cylinder of the sucking pump and leading it into an air-chamber, where it is covered by a valve opening upwards, and using a solid piston without valves, we have the ordinary force-pump. The delivery-pipe enters at the top of the air-chamber and extends downward near to the bottom of the chamber, as shown in Fig. 105. During the downward stroke of the piston the water is forced through the sleeping valve of the air-chamber, compressing the air in the chamber, and the water is forced by the compressed air up the delivery-pipe to the discharge. In both the lifting and force pumps the height to which water can be delivered is limited by the power of the motor and the strength of the pump; while the sucking pump can only be used where the discharge-pipe is at an altitude less than the height of the water barometer.

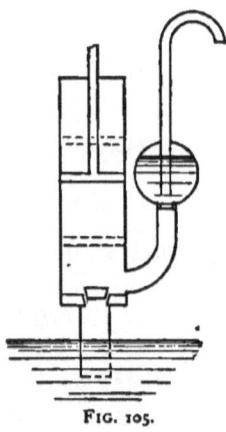

FIG. 105.

276. *Quantity of Energy Expended in Pumping.*—The resistance to be overcome in the upward stroke of the sucking or lifting pump, when friction is neglected, is found thus: If C be the area of the piston, p_a the atmospheric pressure per unit of area, h' the distance of the piston below the discharge, and h_1 its altitude above the reservoir, we have for the pressure downward on the upper surface of the piston,

$$p_a C + \omega C h';$$

and for the pressure upward on the under surface,

$$p_a C - \omega C h_1.$$

Therefore the resistance to be overcome by the motor is the difference of these pressures, or

$$\omega C (h' + h_1) = \omega C h; \quad \ldots \quad (981)$$

h being the height of the discharge above the surface level of the reservoir. As this resistance is constant for a given pump, the work done by the motor in the whole upward stroke is

$$\omega Chp; \qquad \qquad (982)$$

that is, *equal to the quantity of work required to lift a volume of water whose base is the area of the piston, and whose altitude is the vertical distance from the level of the reservoir to the discharge-pipe, through a height equal to the play of the piston.* To this work must be added that consumed by friction in both the upward and downward strokes.

In the force-pump the work of the motor is employed in both strokes, and, since in the upward stroke the sleeping valve in the air-chamber is closed, and in the downward stroke it is open and the other valve closed, the resistance is variable. Using the same notation, we will have for the resistance at any point of the upward stroke at an altitude z above the level of the reservoir,

$$\omega C z ;$$

and for the elementary quantity of work during the ascent,

$$\omega C z dz ;$$

and for the whole work during ascent,

$$\omega C \int_{z_1}^{z'} z \, dz = \omega C \frac{z' + z_1}{2} (z' - z_1) = \omega C \frac{z' + z_1}{2} p; \qquad (983)$$

in which z' and z_1 are the heights of the upper and lower limits of the play of the piston above the level of the reservoir. If the altitude of the middle point of the play be represented by h', the expression for the work done during the upward stroke may be written

$$\omega C h' p.$$

During the downward stroke the variable pressure to be overcome, when the piston is at any point z'' below the discharge, is

$$\omega C z'';$$

and if h_1 be the distance of the middle point of the play below the discharge, we will find, by the same method as above, that the quantity of work done during the downward stroke of the piston is

$$\omega C h_1 p.$$

The total work done during the double stroke, excluding that consumed by friction, is then

$$\omega C (h' + h_1) p = \omega C h p; \quad \ldots \quad (984)$$

that is, *equal to the quantity of work necessary to raise a volume of water, whose base is the area of the piston, and whose altitude is the height of the discharge above the level of the reservoir, through a height equal to the play of the piston.* As this is precisely the same quantity of work required to raise a volume of water whose base and altitude are C and p through a height h, and as this latter volume is the amount discharged during a double stroke of the piston, in either the sucking, lifting or force pump, we see that the motor must supply all the required energy. Without the aid of the atmosphere, the sucking pump could not be used; but it is to be remembered that it does not supply any energy for the accomplishment of the work required to raise water from one level to another.

277. *The Centrifugal Pump.*—For pumping large quantities of water through a small height the centrifugal pump is very advantageous. It consists of a wheel similar to a radial-flow turbine, revolving in a circular chamber. The water enters at the axis, and the centrifugal force developed by the rotation causes the water to flow outward into the annular chamber surrounding the wheel, and thence through a pipe to the point where it is to be discharged.

278. *Double-acting Force-pump.*—Very high power pumps, such as are required to supply large quantities of water at a great height, or to project a stream with a high velocity, as in the case of a fire-engine, are constructed on the principle of the force-pump. The pump is made double-acting, that is, water is discharged into the pipe and air-chamber at each single stroke of the piston, the arrangement of the valves being the same at both ends of the play.

279. *The Hydraulic Ram.*—This is a machine by means of which a quantity of water falling from a height h forces a portion of the water to a greater height h'. It consists of a supply

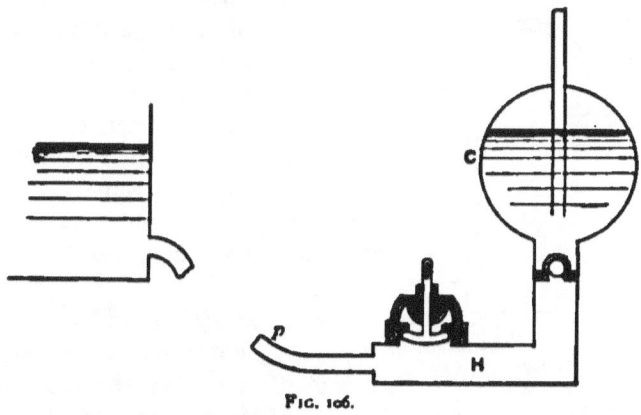

FIG. 106.

reservoir, Fig. 106, which derives its water from a natural stream; a sufficiently long pipe p, which leads the water to the head of the ram; the ram itself, which consists of the head H, and the ram valve opening downward. To the head a discharge-pipe is attached having a valve opening upward, and above and connected with the discharge-pipe is an air-chamber C. The method of operation is briefly this: The weight of the ram valve is greater than the statical pressure of the water on its under side, and being open, the water escapes around the seat of this valve. But when the velocity of the outgoing water attains such a value as to make the pressure on the under side greater than its weight, the valve is lifted and the opening is closed. The sudden stoppage of motion of the water causes a greater pressure on the pipe than existed during the flow, and this pressure opens the valve in the delivery-pipe and forces a portion of the water into the air-chamber. In a short time the pressure on the pipe becomes again equal to the statical, the ram valve falls to its original position, and the flow begins again, and so on, intermittently. Part of the energy of the flowing water is therefore employed in compressing the air in the chamber, which in turn is expended in lifting the water in the delivery-pipe to the discharge. Another part is employed in expanding the supply-pipe, causing a recoil of the water from the head of the ram and a lessened pressure, which thus assists in the fall of the ram valve. The stroke of the ram valve against its seat, being somewhat similar to the butt of a ram, gives the name to this device.

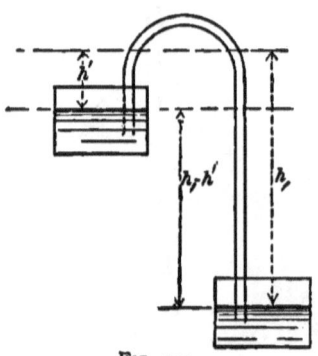

Fig. 107.

280. *The Siphon.*—The siphon, Fig. 107, is a bent tube, its two branches being of unequal length, which is used to convey water or other liquid from a higher to a lower reservoir over an elevation higher than the level of the upper reservoir of the liquid. The height of the intervening elevation above the surface to which the upper level is to be drained must not

exceed the height of liquid which the atmospheric pressure will sustain in vacuo at the place. Eq. (925) is applicable to steady flow through the siphon, and the velocity of discharge is therefore

$$v = \sqrt{2gh} = \sqrt{2g(h_1 - h')}, \quad \ldots \ldots \quad (985)$$

when viscosity, fluid friction and other resistances are neglected.

In the siphon, as in the case of pumps, the atmospheric pressure does no work; and although a siphon could not be used without its assistance, we see that the total work done is wholly due to the transformation of the potential energy of the liquid, stored in the upper reservoir, into an equal quantity of kinetic energy, or $\tfrac{1}{2}Mv^2$ foot-pounds of work.

281. *The Air-pump.*—Any machine designed to diminish the density of airs within enclosed rigid vessels is called an *air-pump*. All such contrivances are based on the fundamental principle that airs are capable of expanding indefinitely, and in direct proportion to the decrease of the external pressure. The essential parts of an air-pump are the *receiver*, the *cylinder and piston*, the *valves* or *stop-cocks*, and the *gauge*.

The receiver is a vessel with rigid sides, properly arranged so that its interior can be connected with the interior of the cylinder. The cylinder and piston in the ordinary air-pump are the most important parts of the machine, for upon their perfection and that of the valves depends the value of the pump. The piston must move air-tight under the external pressure of the atmosphere, and should be of such a form as to leave a minimum space between it and the bottom of the cylinder when it reaches the bottom of its play. The gauge is a barometer or manometer attached to the tube connecting the cylinder and receiver, to indicate the degree of rarefaction attained at any instant.

Let r be the interior volume of the receiver and connecting tube, and c that of the portion of the cylinder comprised between the extreme limits of the play of the piston. Then if communication be opened between the receiver and the cylinder, the piston being supposed at its lower limit, and the piston be moved to the further limit, the air originally in the receiver and communicating tube will, by its expansion, fill the receiver, tube and cylinder. The density of the air δ', in the increased volume will be less than δ, the original density, and by Boyle's law we will have

$$\delta r = \delta'(r+c),$$

or

$$\delta' = \delta \frac{r}{r+c}. \quad \ldots \ldots \quad (986)$$

After closing the communication between the receiver and cylinder, the air in the cylinder can be driven out through another valve into the external atmosphere by returning the piston to its former position. Exhausting again, and representing the new density by δ'', we have

$$\delta' r = \delta''(r+c);$$

whence

$$\delta'' = \delta' \frac{r}{r+c} = \delta \left(\frac{r}{r+c}\right)^2;$$

and after n double strokes the density of the air in the receiver becomes

$$\delta_n = \delta \left(\frac{r}{r+c}\right)^n. \quad \ldots \ldots \quad (987)$$

Hence we see that it is theoretically impossible to expel all the air from the receiver, but it is evident that with a good air-pump it is practically possible to attain so great a degree of exhaustion as to make the residue of air insignificant with respect to the original quantity. The best air-pumps of this kind can-

PUMPS. 355

not reduce the pressure to a less amount than that which corresponds to one or two millimeters of mercury.

282. *The Mercurial Air-pump*.—To obtain the greatest possible degree of exhaustion recourse must be had to the mercurial air-pump, of which there are two forms.

The first, or *Geissler's*, Fig. 108, consists essentially of two communicating vessels of mercury, one of which can be alter-

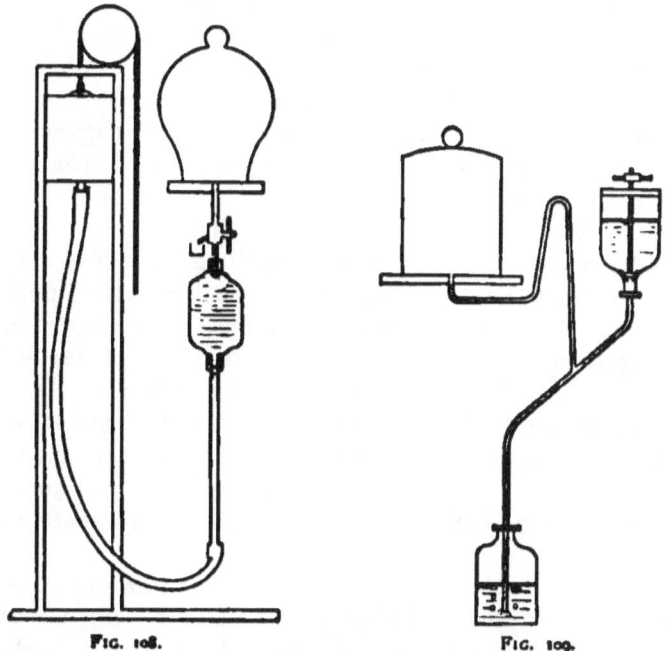

Fig. 108.   Fig. 109.

nately raised above and lowered below the other, so that the liquid may flow from one into the other. The stationary vessel is then the cylinder, and the mercury in the movable vessel acts as a reciprocating moving piston, alternately leaving a space into which the expanding air may flow, and then driving out this air into the external atmosphere.

The second form, or *Sprengel* pump, Fig. 109, consists of a reservoir of mercury from which the liquid flows down a vertical tube to a lower level. Attached to this tube below the reservoir is an inclined tube communicating with the receiver, and through which the expanding air of the receiver enters discontinuously, or in the form of bubbles. The cylinder in this case is practically the vertical tube, and the piston is a series of small mercury cylinders moving continuously down the tube. The highest degree of exhaustion attainable by the mercurial air-pump corresponds to a pressure of .00015 mm., or about 0.0000002 of an atmosphere.

283. *The Atmosphere and Atmospheric Pressure.*—The mass of air which envelops the earth is called the atmosphere. It is mainly a mixture of the two gases, oxygen and nitrogen. The resistance with which it opposes the motion of bodies of small mass and relatively large volume, and the striking effects produced by wind or air in motion, such as the destruction of solidly constructed buildings, uprooting trees, the motion of ships, etc., sufficiently attest its existence. The upper limit of the atmosphere is unknown, but at a distance of 200 miles its density is so slight that we may consider the mass of air beyond this distance to be insignificant. The lower strata of the atmosphere support all above and are compressed by their weight. The atmosphere as a whole is retained as an envelope around the earth by the attraction of the earth for the matter which composes it.

The *atmospheric pressure* on any surface is nothing more than the weight of a column of air, whose base is the surface pressed and whose altitude extends to the upper boundary. If a straight glass tube about 36 inches long, closed at one end, be filled with mercury and inverted so as to allow the open end to dip below the surface of mercury in an open vessel, it will be found, when communication between the tube and vessel is freely made, that after a few oscillations the mercury in the tube will stand at a height of about 30 inches, provided the tube is free from air and watery vapor. The free space in the tube above the mercury contains no known ponderable matter except perhaps some vapor of mercury; the space is called the *Torricellian vacuum*. Equilibrium is then maintained between the air-pressure on an area equal to the cross-section of the tube and the pressure of the liquid column of mercury. The height of the mercury column in the tube will vary with the temperature of the mercury, the latitude of the place, the height above the sea-level and be affected by temporary atmospheric changes. If water be used in the tube instead of mercury, the height of the column will be nearly 34 feet.

284. *The Barometer.*—A Torricellian tube fixed in a permanent position, so that the height of its mercury column can be readily ascertained, is called a *Barometer*, or *weight-measurer of the air*. In order that its indications may be accurate, the mercury should be chemically pure, and the tube should be thoroughly clean and free from air and moisture. To avoid the necessity of correcting for capillarity, the cross-section of the tube should be at least equal to 20 mm. As mercury expands with an increase in temperature, the heights of the mercurial column will differ for different temperatures of the mercury even when the air-pressure remains the same. It is therefore necessary to reduce the observed height to what it would have been at the standard temperature under the same pressure.

Let m represent the coefficient of expansion for mercury, $h_{\circ}$ and h the heights of the mercury column under the same pressure at $0°$ and $t°$ respectively : then we will have

$$h_{\circ}(1 + mt°) = h;$$

whence

$$h_{\circ} = h(1 - mt°), \text{ very nearly.}$$

For the Centigrade scale, m is .0001801; and for the Fahrenheit, .0001001.

The principal forms of the mercurial barometer in use are

the Cistern and Siphon. In the former the level of the mercury in the cistern can be raised or lowered by means of a screw acting on the flexible bottom so as to bring the upper surface of the mercury in the cistern to a fixed origin, from which the height of the column is measured. In the latter the origin of the scale is brought to the level of the mercury in the shorter leg of the siphon. The siphon barometer is usually employed when the barometer is permanently fixed, as in observatories, and the cistern where the instrument must be transported from place to place. The *Aneroid* barometer consists essentially of a thin metallic box from which the air is exhausted. The variations of air-pressure produce changes in the box which are measured on a dial by a suitable mechanism. The apparatus employed to measure the elastic force of any gas or vapor are called *manometers* or *pressure-gauges*, and their indications are usually expressed in terms of that atmospheric pressure which is assumed as a standard.

285. *Standard Atmosphere.*—Let h be the height, and the square centimeter the base, of the mercurial column which measures the atmospheric pressure p, and δ_m the density of mercury: then we have

$$p = h \times 1 \times \delta_m \times g = h \delta_m g. \quad \ldots \quad (988)$$

At Paris the value of g is 980.94 dynes; then if h be $.76^m$, we have at 0° C.,

$$p = 76 \times 13.596 \times 980.94 = 1.0136 \times 10^6 \text{ dynes} \quad . \quad (989)$$

per square centimeter. At Greenwich, g being 981.17 dynes, we have, under the same conditions,

$$p = 76 \times 13.596 \times 981.17 = 1.0138 \times 10^6 \text{ dynes}.$$

Since the density of mercury is a constant, the pressure corresponding to 76 cm. varies slightly, due to a variation in the value of g. Assuming the *standard atmospheric pressure* to be that of 10^6 dynes, we have

$$p = h \times \delta_m \times g = 10^6 \text{ dynes}, \quad \ldots \quad (990)$$

from which we have

$$h = \frac{10^6}{13.596 \times g} \quad \ldots \ldots \quad (991)$$

for the height of the mercury column corresponding to a standard atmosphere at any latitude. Substituting the values of g for Greenwich, Paris, and West Point, we have 74.964, 74.979 and 75.029, for the values of h respectively.

286. *Density of Air.*—Regnault found the density of perfectly dry air at Paris to be 0.0012932 grams per cubic centimeter, corresponding to a pressure of 76 cm. at 0° C. Since this pressure is equal to 1.0136×10^6 dynes, the density of dry air at the standard atmospheric pressure is $0.0012932 \div 1.0136 = 0.0012759$, and at any pressure will therefore be, by Boyle's law,

$$\delta_a = p \times 1.2759 \times 10^{-9}.$$

287. *Height of Homogeneous Atmosphere.*—Let H be the height of a homogeneous fluid column whose density is that of the atmosphere δ_a: then we have, Eq. (988),

$$p = H \delta_a g; \quad \ldots \ldots \quad (992)$$

whence, for dry air at 0° C.,

$$H = \frac{p}{\delta_a g} = \frac{p \times 10^9}{p \times g \times 1.2759} = \frac{7.8376 \times 10^8}{980.6056 \,(1 - .002552 \cos 2\lambda)}$$
$$= 7992^m.61 \,(1 + .002552 \cos 2\lambda); \quad \ldots \ldots \quad (993)$$

from which the value of H can be found for any latitude and, by the application of Charles's law, for any temperature. If the temperature be supposed constant, and the changes in g for variations in the height above the sea-level be disregarded, then H will be constant at the same place for all altitudes, since, Eq. (992),

$$H = \frac{p}{\delta_a g},$$

in which g and the ratio $p \div \delta_a$ are constant by hypothesis.

288. *The Barometric Formula.*—If the atmosphere be supposed in equilibrium, a definite relation must exist between the altitude of any point above the sea-level and the atmospheric pressure at that point. To find this relation, let z be the vertical coordinate at any point, and p the corresponding pressure: then the variation in the pressure dp is due to the weight of the indefinitely thin stratum of air whose altitude is dz; whence we may write

$$-dp = g\delta_a dz = \frac{p}{H}dz, \quad \ldots \ldots \quad (994)$$

or

$$\frac{dz}{H} = -\frac{dp}{p}. \quad \ldots \ldots \ldots \quad (995)$$

Integrating between limits corresponding to the altitudes z_1 and z_2, we have

$$z_2 - z_1 = H \text{ Nap. } \log \frac{p_1}{p_2}; \quad \ldots \ldots \quad (996)$$

the *Barometric Formula*, previously deduced in Art. 260. Substituting the value of H, Eq. (993), and changing to common logarithms by dividing by the common modulus, 0.4342945, we have

$$z_2 - z_1 = 18403^{\text{m}}.65 \ (1 + .002552 \cos 2\lambda) \log \frac{p_1}{p_2}; \quad (997)$$

As the temperature of the air is scarcely ever the same at the two altitudes z_2 and z_1, an approximate value of H is taken, under the supposition that the temperature of the whole body of air between the two stations is a mean of those at the stations. Hence if t_1 and t_2 be the temperatures of the air at the lower and upper stations respectively, we will have

$$z_2-z_1 = 18403^m.65\,(1+.002552\cos 2\lambda)\left(1+\alpha\frac{t_1+t_2}{2}\right)\log\frac{p_1}{p_2}. \quad (998)$$

Replacing the ratio of the pressures by the equal ratio of the barometric heights, reduced to standard temperatures, by the introduction of the factor of expansion for mercury, we have finally

$$z_2-z_1 = 18403^m.65\,(1+.002552\cos 2\lambda)\left(1+.003665\frac{t_1+t_2}{2}\right)$$
$$\left(\log\frac{h_1}{h_2}\cdot\frac{1}{1+(T_1-T_2).0001801}\right). \quad (999)$$

To pace this in a form convenient for logarithmic computation, let

$$18403^m.65\left(1+.003665\frac{t_1+t_2}{2}\right) = A,$$
$$1+.002552\cos 2\lambda = B,$$
$$\frac{1}{1+(T_1-T_2).00018} = C.$$

Then Eq (999) may be written

$$z = z_2-z_1 = AB\log\frac{Ch_1}{h_2} = AB(\log C + \log h_1 - \log h_2);$$

and taking the logarithms of both members, we have

$$\log z = \log A + \log B + \log(\log C + \log h_1 - \log h_2). \quad (1000)$$

The values of A for all possible values $t_1 + t_2$, of B for all latitudes, and of C for all values of $T_1 - T_2$, may be previously determined and formed into tables for ready reference.

289. The barometric formula is deduced under several conditions which are scarcely ever wholly satisfied in the actual use of the barometer. In the first place, the air is supposed to be in equilibrium, and this condition requires that observations should only be made when there is no wind. In the next place, the observations are supposed to be made simultaneously at the upper and lower stations, which is only practicable when two instruments are used. For a single observer three observations are necessary, two at one station and the intermediate observation to be taken at the other station at a time midway between the first and last observations. The sequence to be followed by the observer is generally this: After mounting the barometer on its stand at the station, it is brought to the level of its zero point; the temperature of the air is taken by means of a detached thermometer previously placed in the suitable place; the height h_1 of the barometer is read off from the scale, and the temperature T_1 of the mercury is taken from the thermometer found attached to the barometer tube. A mean of the observations at the first station properly reduced is taken as the correct data which would be simultaneous with those observed at the other station, and these mean values take the place of t_1, T_1, and h_1, in the formula.

APPENDIX.

TABLE I.

DENSITY AND SPECIFIC GRAVITY.

1 cubic foot of distilled water at 39.2° F. weighs 62.425 lbs.
1   "   "   "   "   62° F.   "   62.355 lbs.

SOLIDS.		
Substances.	Density. Water at 62° F. = 1.	Weight of Cubic Foot in Pounds.
Metals*—		
Aluminium.........................	2.55– 2.65	159– 165
Antimony..........................	6.66– 6.74	415– 420
Brass, cast........................	7.8 – 8.4	486– 524
" rolled	8.4 – 8.5	524– 530
" wire	8.54	533
Bronze, gun-metal..................	8.45– 8.85	527– 552
Copper, cast	8.6 – 8.8	536– 549
" wrought...................	8.8 – 9.0	549– 561
Gold...............................	19.3 –19.6	1203–1222
Iron, cast..........................	6.9 – 7.4	430– 461
" " gun-metal..................	7.25– 7.4	452– 461
" wrought...................	7.6 – 7.9	474– 493
Lead...............................	11.3 –11.47	705– 715
Platinum	19.5 –22.0	1216–1372
Silver..............................	10.5	655
Steel...............................	7.8 – 7.9	486– 493
" gun-metal.....................	7.84– 7.88	489– 491
Tin.................................	7.2 – 7.5	449– 469
Zinc................................	6.8 – 7.2	424– 449
Wood†— The density of a single variety varies, but will seldom differ either way from the tabular values by more than $\frac{1}{16}$. Those given are average values for dry, well-seasoned woods. Green wood weighs $\frac{1}{4}$ to $\frac{1}{3}$ more, and ordinary building timber, tolerably seasoned, about $\frac{1}{8}$ more.		
Ash................................	.6 – .7	37– 44

* Mostly from Trautwine's "Engineer's Pocket-Book."
† Mostly from "Ordnance Manual" and Trautwine's "Engineer's Pocket-Book."

TABLE I.—Continued.

SOLIDS—Continued.

Substances.	Density. Water at 62° F. = 1.	Weight of Cubic Foot in Pounds.
Wood—*Continued.*		
Beech...................................	.7	44
Chestnut................................	.5 – .6	31– 37
Cypress.................................	.55	34
Ebony...................................	1.2	75
Elm.....................................	.6 – .7	37– 44
Hickory.................................	.8 – .9	50– 56
Hemlock.................................	.45	28
Lignum-vitæ.............................	1.3	81
Mahogany, Honduras......................	.55	34
" Spanish...................	.85	53
Maple...................................	.6 – .7	37– 44
Oak, white..............................	.7 – .8	44– 50
" live......................	1.0	62
" other varieties...........	.65 – .8	41– 50
Pine, pitch.............................	.7 – .8	44– 50
" yellow....................	.6	37
" white.....................	.4	25
Poplar..................................	.45	28
Spruce..................................	.45	28
Walnut, black...........................	.6	37
Miscellaneous*—		
Asphaltum...............................	1.0 –1.8	62–112
Basalt..................................	2.8 –3.	175–187
Brick...................................	1.5 –2.5	94–156
Charcoal, soft to hard woods............	.25 – .6	16– 37
" for gunpowder.............	.38	24
Clay, dry...............................	1.9	118
Coal, anthracite........................	1.3 –1.8	81–112
" " broken...................	....	52– 60
" bituminous................	1.2 –1.5	75– 94
" " broken...................	....	47– 56
" lignite....................	1.1 –1.25	69– 78
Earth, common, moderately rammed........	1.5	94
" mean of the globe, about..	5.66	....
Glass, crown, average...................	2.5	156
" flint, average............	3.	187
" green, average............	2.7	168
" plate, average............	2.7	168

* Mainly from Trautwine's "Engineer's Pocket-Book."

TABLE I.—*Continued.*

SOLIDS—*Continued.*

Substances.	Density. Water at 62° F. = 1.	Weight of Cubic Foot in Pounds.
Miscellaneous—*Continued.*		
Gneiss........	2.6 –2.8	162–175
Granite.......	2.5 –2.9	156–181
Gunpowder, press-cake.......	1.70 –1.85	106–115
" ordinary grained.......	.875– .900	55– 56
Ice.......	.92	57
India-rubber.......	.95	59
Ivory.......	1.8	112
Limestone, marble.......	2.65 –2.85	165–178
" common building.......	2.4 –2.9	150–181
Nitre, crystallized.......	1.9	118
Quartz.......	2.63	165
Sand, dry to wet.......	1.5 –2.	94–125
Sandstone, building.......	2.1 –2.7	131–168
Slate.......	2.7 –2.9	168–181
Sulphur.......	2.	125
Wax, various kinds.......	.9 –1.	56– 62

LIQUIDS.

Acid, hydrochloric, muriatic, sat. sol....	1.21	75.5
" nitric, concentrated.......	1.5	93.5
" sulphuric, concentrated.......	1.84	114..7
Alcohol, absolute.......	.795	49.6
" proof spirit.......	.92	57.3
Ether, sulphuric, common.......	.72	44.9
Glycerine.......	1.27	79.2
Mercury.......	13.6	848.0
Nitro-glycerine.......	1.6	99.8
Oil, illuminating.......	.8	49.9
" linseed.......	.94	58.6
" olive.......	.92	57.3
" ("spirits") of turpentine.......	.87	54.2
Water, distilled.......	1.	62.4
" sea.......	1.027	64.0

TABLE I.—Continued.

GASES.

Air, dry, at 60° F. and 30 in. Bar., density $= \dfrac{1}{813.88} = .001229$ ⎫
" " " 32° F. " 30 " " " $= \dfrac{1}{770.89} = .001279$ ⎬ Times maximum density of water.
⎭

	Density of Hydrogen = 1.	Density of Air = 1.	Weight of Cubic Foot 60° F., 30". Ozs. avoir.	Weight of Cubic Meter 0° C., 76 cm. Grams.
Air, dry....................	14.422	1.	1.226	1293
Carbonic acid, CO_2...........	22	1.525	1.870	1973
" oxide, CO...........	14	.970	1.190	1255
Coal gas....................	4.76-5.77	.33-.40	.40-.49	427-517
Hydrogen...................	1	.069	.085	89.7
Marsh gas, CH_4..............	8	.555	.680	717
Nitrogen....................	14	.970	1.190	1255
Olefiant gas, C_2H_4...........	14	.970	1.190	1255
Oxygen.....................	16	1.109	1.360	1435
Steam (ideal)................	9	.624	.765	807
" at 212° F..............	....	....	.592	624

TABLE II.

THE METRIC SYSTEM.

The metric system of weights and measures is founded on the *meter* as a unit of length. The units of the system are as follows:

Length : The *Meter* = length of standard bar preserved at Paris.
Area: The *Are* = 100 square meters.
Volume: The *Stere* = 1 cubic meter.
Capacity: The *Liter* = 1 cubic decimeter.
Mass and Weight: The *Gram* = the mass or weight of 1 cubic centimeter of distilled water at the temperature of maximum density.

It is a *decimal system.* The prefixes denoting multiples are derived from the Greek, and are: *deka*, ten; *hecto*, hundred; *kilo*, thousand; and *myria*, ten thousand. Those denoting sub-multiples are taken from the Latin, and are: *deci*, tenth; *centi*, hundredth; and *milli*, thousandth.

The following table includes all the measures of the system in use:

No. of the Unit.	Length.	Area.	Volume.	Capacity.	Mass and Weight.
10000	Myriameter.				Myriagram.
1000	Kilometer.				Kilogram, kg.
100	Hectometer.	Hectare, ha.		Hectoliter, hl.	Hectogram.
10	Dekameter.		Dekastere.	Dekaliter, dal.	Dekagram.
1	Meter, m.	Are, a.	Stere, s.	Liter, l.	Gram, g.
.1	Decimeter, dm.	Declare.	Decistere.	Deciliter, dl.	Decigram, dg.
.01	Centimeter, cm.	Centiare.		Centiliter, cl.	Centigram, cg.
.001	Millimeter, mm.				Milligram, mg.

The *are* and its derivatives are used only for *land measure.* In other cases area is expressed in terms of the square whose side is a measure of length— e.g., square meter, m^2; square centimeter, cm^2, etc.

The *stere* is rarely used except in measuring *firewood.* In other cases a cube whose edge is a unit of length is used—e.g., cubic meter, m^3; cubic dekameter, dm^3, etc. Cubic dekameter, cubic hectometer, etc., are not used.

A *Mikron*, μ, = .001 mm.
A *Tonne*, *t*, or millier, = 1000 kg.
A *Metric Quintal*, *q*, = 100 kg.

TABLE III.

FOR CONVERTING METRIC INTO UNITED STATES MEASURES, AND VICE VERSA.

The value of the meter in inches, and that of the gram in grains, are those in use in the United States Coast and Geodetic Survey, 1887.

Metric to U. S.	Number.	Logarithm.		Logarithm.	Number.	U. S. to Metric.
Inches in 1 centimeter.	0.39370428	1̄.5951701		0.4048299	2.539978	Centimeters in 1 inch.
Feet in 1 meter.	3.280869	0.5159889		1̄.4840111	0.3047973	Meters in 1 foot.
Yards in 1 meter.	1.093623	0.0388676		1̄.9611324	0.9143918	Meters in 1 yard.
Miles in 1 kilometer.	0.6213769	1̄.7933551		0.2066449	1.609329	Kilometers in 1 mile.
Square inches in 1 sq. centimeter.	0.1550030	1̄.1903403		0.8096597	6.451484	Square centimeters in 1 sq. inch.
Square feet in 1 square meter.	10.76410	1.0319778		2.9680222	0.0929038	Square meters in 1 square foot.
Square yards in 1 square meter.	1.196011	0.0777353		1̄.9222647	0.8361126	Square meters in 1 square yard.
Acres in 1 hectare.	2.471098	0.3928899		1̄.6071101	0.4046985	Hectares in 1 acre.
Cubic inches in 1 cubic centimeter.	0.06102537	2̄.7855104		1.2144896	16.38663	Cubic centimeters in 1 cubic inch.
Cubic feet in 1 cubic meter.	35.31561	1.5479667		2.4520333	0.02831609	Cubic meters in 1 cubic foot.
Cubic yards in 1 cubic meter.	1.307985	0.1166029		1̄.8833971	0.7645345	Cubic meters in 1 cubic yard.
Cords (128 cu. ft.) in 1 stere.	0.275902	1̄.4407567		0.5592433	3.624460	Steres in 1 cord.
Quarts (standard liquid) in 1 liter.	1.056716	0.0239584		1̄.9760416	0.9463279	Liters in 1 quart (liquid).
Gallons (standard liquid = 231 cu. in.) in 1 dekaliter.	2.641791	0.4218984		1̄.5781016	0.3785311	Dekaliters in 1 gallon.
Quarts (standard dry) in 1 liter.	0.908107o	1̄.9581371		0.0418629	1.101192	Liters in 1 quart (dry).
Bushels (standard struck = 2150.42 cu. in.) in 1 hectoliter.	2.837835	0.4529871		1̄.5470129	0.3523813	Hectoliters in 1 bushel.

APPENDIX.

Grains in 1 gram.	15.4326376	1.1884325	2.8115673	0.0647087	Grams in 1 grain.
Ounces (avoir.) in 1 gram.	0.0352739	2.5474556	1.4535456	28.34951	Grams in 1 ounce (avoir.).
Pounds (avoir.) in 1 kilogram.	2.2046339	0.3433344	1.6366356	0.4535922	Kilograms in 1 pound (avoir.).
Pounds (avoir.) in 1 tonne (millier).	2204.623	3.3433344	4.6366356	0.00045359	Tonnes (milliers) in 1 pound (avoir.).
Tons (2240 lbs.) in 1 tonne (millier).	0.9842067	1.9936064	0.0069136	1.016046	Tonnes (milliers) in 1 ton (2240 lbs.).
Foot-pounds in 1 kilogram-meter.	7.23300	0.8593233	1.1406707	0.1389337	Kilogram-meters in 1 foot-pound.
Foot-tons in 1 tonne-meter.	3.22954	0.5090753	1.0099247	0.309688	Tonne-meters in 1 foot-ton.
Fahrenheit degrees in 1 centigrade degree.	1.8	0.2552725	1.7447275	0.5555555	Centigrade degrees in 1 Fahrenheit degree.
Fahr.-pound heat-units in 1 cent.-kilogram heat-units.	3.9683222	0.5986069	1.4013931	0.2519057	Cent.-kilogram heat-units in 1 Fahr.-pound heat-unit.
Pounds-to-the-inch in 1 kilogram-to-the-centimeter.	5.599693	0.7481643	1.2518357	0.1785812	Kilograms-to-the-centimeter in 1 pound-to-the-inch.
Pounds-to-the-foot in 1 kilogram-to-the-meter.	0.6719631	1.8273455	0.1726545	1.488177	Kilograms-to-the-meter in 1 pound-to-the-foot.
Tons-to-the-foot in 1 tonne-to-the-meter.	0.9999835	1.4779975	0.5009005	3.333515	Tonnes-to-the-meter in 1 ton-to-the-foot.
Pounds-to-the-sq.-inch in 1 kilogram-to-the-sq.-centimeter.	14.22329	1.1529941	2.8470059	0.0703080	Kilograms-to-the-sq.-centimeter in 1 pound-to-the-sq. inch.
Pounds-to-the-sq.-foot in 1 kilogram-to-the-sq.-meter.	0.2048186	1.3113566	0.6886434	4.883512	Kilograms-to-the-sq.-meter in 1 pound-to-the-square-foot.
Tons-to-the-sq.-foot in 1 tonne-to-the-sq.-meter.	0.09143418	2.9610086	1.0389914	10.9363	Tonnes-to-the-sq.-meter in 1 ton-to-the-sq.-foot.
Pounds-to-the-cubic-inch in 1 kilogram-to-the-cubic-centimeter.	36.12634	1.5578340	2.4421760	0.0276863	Kilograms-to-the-cubic-centimeter in 1 pound-to-the-cubic-inch.
Pounds-to-the-cubic-foot in 1 kilogram-to-the-cubic-meter.	0.06246231	2.7953677	1.2046323	16.01889	Kilograms-to-the-cubic-meter in 1 pound-to-the-cubic-foot.

TABLE IV.

GRAVITY.

g = Acceleration due to gravity, in feet per second.
L = Length of simple seconds pendulum, in feet.
λ = Latitude.
h = Height above sea-level, in feet.
$g = 32.173 - 0.0821 \cos 2\lambda - 0.000003 h.$*
$L = 3.2597 - 0.0083 \cos 2\lambda - 0.0000003 h.$*

Values of $g = 32.173 - 0.0821 \cos 2\lambda$ and $L = 3.2597 - 0.0083 \cos 2\lambda$.

Latitude.	g.	L.
0°	32.091 f. s.	3.2514 f.
5	32.092	3.2515
10	32.096	3.2519
15	32.102	3.2525
20	32.110	3.2533
25	32.120	3.2544
30	32.132	3.2556
35	32.145	3.2569
40	32.159	3.2583
45	32.173	3.2597
50	32.187	3.2611
55	32.201	3.2625
60	32.214	3.2638
65	32.226	3.2650
70	32.236	3.2661
75	32.244	3.2669
80	32.250	3.2675
85	32.254	3.2679
90	32.255	3.2680

The value of g is affected to some extent by the character and arrangement of the local geological strata. The variation from the tabular value may be as great as 10 units of the last place of figures, but rarely exceeds 5 of these units.

* Encyclopædia Britannica, art. Gravitation.

TABLE V.

FRICTION.*

Substances.	Angle of Repose. °	Coefficient of Friction. f	$\dfrac{f}{\sqrt{1+f^2}}$
Wood on wood, dry................	14 –26.5	.25–.5	.24–.45
" " " soaped.	11.5	.2	.20
Wood on metals, dry.......	11.5–31.	.2 –.6	.20–.51
Metals on oak, dry................	26.5–31.	.5 –.6	.45–.51
" " " soaped.............	11.5	.2	.20
Metals on elm, dry................	11.5–14.	.2 –.25	.20–.24
Hempen cord on oak, dry..........	26.5	.5	.45
" " " " wet..........	16.5	.3	.29
Metals on metals, dry	8.5–11.5	.15–.2	.15–.20
" " " wet.	16.5	.3	.29
Smooth surfaces, occasion'y greased	4. – 4.5	.07–.08	.07–.08
" " continually "	3.	.05	.05
" " best results.......	1.7	.03	.03

These values are for low velocities and pressures at ordinary temperatures. The coefficient for smooth metal bearings, well oiled, varies somewhat with the pressure and velocity, being generally less than the above. It also varies considerably with the temperature, which affects the lubricant. In favorable cases it has been as low as .002 [Thurston].

---

* Mostly from Rankine's " Rules and Tables," and Trautwine's " Engineer's Pocket-Book."

TABLE VI.

STIFFNESS OF CORDAGE FOR WHITE AND TARRED ROPE.

Morin's Formulas.*

$$S = \frac{K + IW}{D} = \frac{n}{D}(0.002148 + 0.001772n + 0.001191\,W) \text{ for white rope;}$$

$$K = n(0.002148 + 0.001772n), \text{ and } I = n(0.001191).$$

$$S = \frac{K + IW}{D} = \frac{n}{D}(0.01054 + 0.0025n + 0.001372\,W) \text{ for tarred rope;}$$

$$K = n(0.01054 + 0.0025n), \text{ and } I = n(0.001372).$$

Values of K and I in Lbs., for Rope Wound on Axle 1 Foot in Diameter.

n, No. of Yarns.	Ordinary White Rope.			Tarred Rope.		
	Circumference in inches = .4524 $\sqrt{n}$.	Natural Stiffness. K in lbs.	Stiffness due to Tension of 1 lb. I.	Circumference in inches = .5378 $\sqrt{n}$.	Natural Stiffness. K in lbs.	Stiffness due to Tension of 1 lb. I.
6	1.10	0.07668	0.007146	1.32	0.15324	0.008232
9	1.36	0.16286	0.010719	1.61	0.29736	0.012348
12	1.57	0.28094	0.014292	1.86	0.48648	0.016464
15	1.74	0.43092	0.017865	2.08	0.72060	0.020580
18	1.92	0.61279	0.021438	2.28	0.99972	0.024696
21	2.08	0.82656	0.025011	2.46	1.32384	0.028812
24	2.21	1.07222	0.028584	2.64	1.69296	0.032928
27	2.35	1.34978	0.032157	2.80	2.10708	0.037044
30	2.47	1.65924	0.035730	2.95	2.56620	0.041160
33	2.60	2.00059	0.039303	3.09	3.07032	0.045276
36	2.72	2.37204	0.042876	3.23	3.61944	0.049392
39	2.84	2.77888	0.046449	3.36	4.21356	0.053508
42	2.94	3.21602	0.050022	3.48	4.85268	0.057624
45	3.05	3.68496	0.053595	3.61	5.53680	0.061740
48	3.17	4.18579	0.057168	3.73	6.26592	0.065856
51	3.26	4.71852	0.060741	3.84	7.04004	0.069972
54	3.35	5.28314	0.064314	3.95	7.85916	0.074088
57	3.45	5.87966	0.067887	4.07	8.72328	0.078204
60	3.54	6.50808	0.071460	4.17	9.63240	0.082320

* Adapted from Morin's formulas, "Cours de Mécanique," vol. ii., Dulos, pp. 193, 194.

TABLE VII.

*RELATIVE DENSITY OF WATER AT DIFFERENT TEMPERATURES.**

Temp. C.°	Relative Density.	Temp. C.°	Relative Density.	Temp. C.°	Relative Density.
0	.999871	13	.999430	35	.99418
1	.999928	14	.999299	40	.99235
2	.999969	15	.999160	45	.99037
3	.999991	16	.999002	50	.98820
4	1.000000	17	.998841	55	.98582
5	.999990	18	.998654	60	.98338
6	.999970	19	.998460	65	.98074
7	.999933	20	.998259	70	.97794
8	.999886	22	.997826	75	.97498
9	.999824	24	.997367	80	.97194
10	.999747	26	.996866	85	.96879
11	.999655	28	.996331	90	.96556
12	.999549	30	.995765	100	.95865

* Units and Physical Constants.—Everett.

www.ingramcontent.com/pod-product-compliance
Lightning Source LLC
Chambersburg PA
CBHW032021220426
43664CB00006B/321